AF389156

Biosensors: 10th Anniversary Feature Papers

Biosensors: 10th Anniversary Feature Papers

Editors

Teresa A. P. Rocha-Santos
João P. da Costa

MDPI • Basel • Beijing • Wuhan • Barcelona • Belgrade • Manchester • Tokyo • Cluj • Tianjin

Editors
Teresa A. P. Rocha-Santos
University of Aveiro
Portugal

João P. da Costa
University of Aveiro
Portugal

Editorial Office
MDPI
St. Alban-Anlage 66
4052 Basel, Switzerland

This is a reprint of articles from the Special Issue published online in the open access journal *Biosensors* (ISSN 2079-6374) (available at: https://www.mdpi.com/journal/biosensors/special_issues/10_years).

For citation purposes, cite each article independently as indicated on the article page online and as indicated below:

LastName, A.A.; LastName, B.B.; LastName, C.C. Article Title. *Journal Name* **Year**, *Volume Number*, Page Range.

ISBN 978-3-0365-3571-5 (Hbk)
ISBN 978-3-0365-3572-2 (PDF)

Contents

About the Editors

Teresa A. P. Rocha-Santos graduated in Analytical Chemistry (1996), obtained a PhD in Chemistry (2000) and an Aggregation in Chemistry (2018) at the University of Aveiro (UA), Portugal. Presently, she is a Principal Researcher at Centre for Environmental and Marine Studies (CESAM) and the Department of Chemistry of UA (since 2014) and Vice-coordinator of the CESAM Research Centre. Her research concentrates on the development of new analytical methodologies fit for purpose and on the study of emerging contaminants (such as microplastics). She has published 170 scientific papers and has an h-index of 43. She is the editor of seven books. She is a member of the editorial board of Current Opinion in Environmental Science and Health, Data In Brief, Science of the Total Environment, Sensors, Molecules, and Associate Editor of Euro-Mediterranean Journal for Environmental Integration, as well as being an Associate Editor of the Journal of Hazardous Materials and Co-Editor in Chief of Journal of Hazardous Materials Advances.

João P. da Costa graduated with a degree in Biotechnological Engineering in 2005. Soon thereafter, he successfully completed his Master's degree in Medical Diagnostics. He was then a visiting scientist at a biotech-based pharmaceutical company in the U.S.A., and, in 2014, he concluded his PhD in Environmental Chemistry. He has expertise in a wide range of fields of research, including biotechnology-based solutions for environmental remediation, as well as biosensing applications for water monitoring. In recent years, João has gradually shifted his focus towards the issue of microplastic pollution from different perspectives, including at the detection, quantification, mitigation and regulatory levels. He is the author or co-author of 72 peer-reviewed publications and has a h-index of 25. He is also the co-inventor of a patent for a biosensor for the detection and quantification of alkylphenols in aquatic samples.

Preface to "Biosensors: 10th Anniversary Feature Papers"

Defined as devices that produce a measurable signal that is proportional to the concentration of the targeted analyte, biosensors incorporate a biological sensing element and measure signals that derive from biological interactions. Easy, fast, low cost and both highly sensitive and selective, biosensors have contributed to significant advances in next-generation medicine, including individualized medicine and the ultrasensitive point-of-care detection of markers for multiple diseases. Environmental, water and food quality monitoring, as well as drug delivery are also fields that have benefited from the exponential development in biosensors achieved in the last decade. With the advances observed in nanotechnology, research and development in biosensing applications have become open and expanded the multidisciplinary nature of biosensors, as these materials provide the possibility of improving their performance, allowing for an enhanced power of detection and quantification through the control of size and morphology.

This book marks the 10 year anniversary of Biosensors, a period during which the journal published more than 50 Special Issues and 550 papers, welcoming the breakthroughs and innovations in biosensors developed for applications in the areas of food, health and environment, as well as security and defense, topics that were covered in this Special Issue, now published as a book.

Teresa A. P. Rocha-Santos, João P. da Costa
Editors

 *biosensors*

Review

Biosensors for Antioxidants Detection: Trends and Perspectives

Melinda David [1], Monica Florescu [1] and Camelia Bala [2,3,*]

[1] Department of Fundamental, Prophylactic and Clinical Disciplines, Faculty of Medicine, Transilvania University of Brasov, Str. Universitatii no. 1, 500068 Brasov, Romania; melinda.david@unitbv.ro (M.D.); florescum@unitbv.ro (M.F.)

[2] Laboratory for Quality Control and Process Monitoring, University of Bucharest, 4-12 Elisabeta Blvd., 030018 Bucharest, Romania

[3] Department of Analytical Chemistry, University of Bucharest, 4-12 Elisabeta Blvd., 030018 Bucharest, Romania

* Correspondence: camelia.bala@chimie.unibuc.ro

Received: 29 July 2020; Accepted: 29 August 2020; Published: 1 September 2020

Abstract: Herein we review the recent advances in biosensors for antioxidants detection underlying principles particularly emphasizing advantages along with limitations regarding the ability to discriminate between the specific antioxidant or total content. Recent advances in both direct detection of antioxidants, but also on indirect detection, measuring the induced damage on DNA-based biosensors are critically analysed. Additionally, latest developments on (bio)electronic tongues are also presented.

Keywords: antioxidants; biosensor; nanomaterials

---

1. Introduction

Antioxidants (AOx) play an important role, since they represent a defense system of all aerobe organisms, especially in the case of humans, where as a consequence of the metabolic and physiological processes, unstable reactive substances are generated as byproducts [1]. These unstable substances are referred to as reactive oxygen/nitrogen species (ROS/RNS), and they are molecules containing oxygen or nitrogen, with one or more unpaired electrons, making them very reactive. Some of the most important oxygen-derived molecules are the radicals of hydroxyl (-OH), singlet oxygen (O_2) and superoxide anion (O_2^-) and hydrogen peroxide (H_2O_2) as non-radical. Nitrogen-derived species are nitric oxide (NO), peroxynitrite (HNO_3^-) or dinitrogen dioxide (N_2O_2) [2]. An increase of reactive oxygen species can overwhelm the natural antioxidant system of the organism leading to oxidative stress (OS). An increasing number of medical studies correlate the presence of OS to various disorders and medical conditions caused by damage inflicted to healthy cells [3]. For a variety of cardiovascular diseases, including strokes, OS has been, at least partially, viewed as one common etiology, with an increased ROS production of the organism [4]. An abnormal oxidation status has further been linked with chronic diseases such as diabetes [5] and neurological diseases such as Alzheimer's [6,7]. In the case of cancer, it has already been proven that the level of ROS is increased [8,9].

Antioxidants have the role of reducing the adverse effects caused by ROS or RNS, and can herewith be divided in two categories. The first category consists of primary, or so-called chain-breaking AOx (e.g., vitamins, carotenes, phenols), which inhibit the oxidation of biomolecules; the second category comprises AOx that prevent ROS/RNS formation. The essential characteristic of an AOx is its property to donate the hydrogen from its active hydroxyl group (A-OH) in order to generate more stable radicals [10]. In more detail, AOx use two major mechanisms to deactivate radicals: hydrogen atom transfer (HAT), mentioned above, and single electron transfer (SET). Through the SET mechanism, the AOx transfers one electron (A-OH$^+$) in order to reduce any compound, being followed by deprotonation in solution [11]. Herewith, the distinction between AOx and phenolic compounds is also a question of oxidation potential, as thoroughly described by Buratti et al. [12]. The antioxidant capacity of phenols depends on their redox properties, which depend on the presence of a phenolic aromatic ring with hydroxyl substituents in their chemical structure. It is well known that anodic peaks at low oxidation potentials (below ~ 0.5 V) occur for compounds with significant antioxidant activity [13], while, at potentials greater than +0.6 V, all phenolic compounds will be oxidized [12]. Some major benefits of AOx are described in the literature. Vitamin C reduces the incidence of degenerative and chronic diseases [14], while polyphenols prevent cardiovascular diseases and show anti-viral and anti-inflammatory properties [15]. Since many fruits, vegetables and medicinal herbs are known for their high content of AOx, proper alimentation and food supplements are often advised. The antioxidant properties of different plants are not only claimed by the pharmaceutical industry, but also the food and cosmetic industry. Herewith, detection and quantification of AOx is of great importance.

Given the increased interest in AOx detection from various resources (foods, supplements, plants) various methods can be employed. Most analytical methods focus on in vitro determination of the antioxidant capacity (AC) following a competitive or noncompetitive reaction, which can be correlated with the anodic area of cyclic voltammetries and/or the electrochemical index using electrochemical measurements [16,17]. In the case of competitive reactions, a competing target molecule is required to compete with the AOx for the reactive species. A good example for this mechanism is the classical technique of chemiluminescence, where the reaction of ROS with chemiluminescence reagents results in a species in an excited state, capable of emitting light. When AOx are added to the solution, and they react with the initiating reactive species, light emission will be inhibited [16]. In the case of non-competitive reactions, AOx compounds interact directly with ROS. Other classical techniques which offer a complex chemical composition analysis of various AOx-containing compounds include chromatographic methods, such as high-performance liquid chromatography used as such [18] or coupled with a 2,2-diphenyl-1-picrylhydrazyl (DPPH) free radical reaction system [19]. Spectroscopic methods are also widely used. Raman spectroscopy allows a fast quality check [20], while Fourier transform infrared spectroscopy (FT-IR) allows the determination of both phenolic content and AC for some rice varieties [21]. Such classical techniques are elaborate, time-consuming, require specific chemical reagents or solvents, and mostly need specialized personnel. To overcome these shortcomings, electroanalytical methods based on sensors and biosensors have been attracting increasing attention and were proven to be more rapid (real-time analysis). Both can be successfully incorporated with the "lab on a chip" technique, offering flexibility, sensitivity and specificity towards the analysed compounds.

The importance of biosensors keeps increasing, since they allow the integration of innovative materials to increase their performance in terms of sensitivity and specificity. The sensitivity of a biosensor depends on the type of transducer (electrochemical, optical) and the technique used to immobilize or functionalize various nanomaterials, polymers that amplify the output signal. Selectivity and specificity depend on the choice of used materials and specific recognition elements such as enzymes or DNA [22].

Although most biosensor configurations focus on the evaluation of the total antioxidant capacity (TAC), this review focuses on biosensors which discriminate among antioxidant species, classes of phenolic compounds, flavonoids or even specific antioxidants such as rutin. Most described biosensors are used for analysis in food or health-related products, but are not limited to these industries. Most reported biosensors are electrochemical ones (amperometric and voltametric), but colorimetric or luminescent assays and optical biosensors are also taken into consideration. A few (bio)electronic tongues are also presented. The described biosensors focus on both direct detection of certain AOx, but also on indirect detection, measuring the induced damage on DNA-based biosensors. Last, a few sensor configurations are also worth mentioning due to their innovation and analytical performances.

What highlights the ability of a biosensor to discriminate among phenolic content, AC or a specific AOx compounds lies mainly on the specific behaviour of the (bio)materials (e.g., enzymes, nanomaterials) used for the biosensor architecture. Additionally, the different electrochemical mechanisms allow the identification of each electroactive compound in complex natural samples [16,22]. Enzymes such as Laccase or Tyrozinase are known to favor the detection of phenolic content. Herewith, Tyrozinase oxidizes monophenols and o-diphenols to their corresponding quinone in a two-electron process, whereas Laccase catalyzes the oxidation of several aromatic substrates in a one-electron process. Addition of nanoparticles such as gold nanoparticles, enhance the enzymatic catalytic activity towards specific compounds such as catechol. Cyclodextrins (a family of oligosaccharides) also act as molecular receptors for phenolic substances. The overall antioxidant capacity can be assessed using the principle of inflicted oxidative damage upon DNA or proteins. Thus, the addition of a specific antioxidant or a mixture of antioxidant compounds (including phenolic compounds) in the presence of ROS, will decrease the effect of oxidative damage, allowing the quantification of AC.

2. Classification of Antioxidant Species

Several criteria, such as activity, solubility, size, kinetics and occurrence, are used to classify antioxidants [23]. The classification and schematic representation of phenolic compounds is shown in Figure 1.

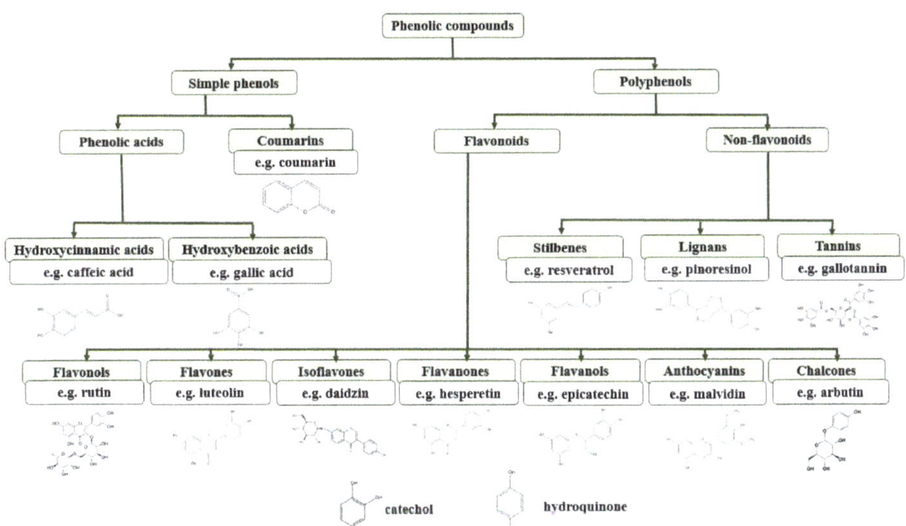

Figure 1. Classification and schematic representation of phenolic compounds.

Most papers in the literature focus on the detection of polyphenols or phenolic compounds due to their health benefits. Polyphenols are usually naturally occurring compounds, found in various plants, and contain multiple functionalities [24]. There are also synthetic phenolic compounds like butylated hydroxyl anisole (BHA), butylated hydroxyltoluene (BHT), and tertiary butyl hydroquinone (TBHQ), for whose detection special enzymatic biosensors have been developed [25]. The activity of polyphenols has been shown to prevent cardiovascular diseases, but also has important anti-inflammatory and anti-viral properties [15]. Polyphenols, alongside vitamins and carotenoids, represent the main classes of dietary AOx. Several classes of polyphenolic compounds can be found in fruits like grapes, apple, pear, cherries and berries [26], or in herbs like lavender [27] and tea [28].

Flavonoids, a widely explored class of polyphenols, are mainly responsible of the colour pigmentation of flowers, fruits and leaves [29]. High contents of flavonols can be found in grapes [30] or walnuts [31]. Vitamins also interrupt free radical chain reactions, and vitamin C (vit C) can be found in various plants [14] and is often consumed as a dietary supplement. There are quite a few ascorbate biosensors to be found in literature, but also vitamins such as vitamins E and D are of interest. A very interesting aspect of biosensors based on ascorbate oxidase is the fact that they enable discrimination between phenols and vit C contribution to the AC of the analysed compound. A review [32] reports a few papers by the group of P.A. Serra in which they highlight the advantages of ascorbate-based biosensors in terms of oxidizing the substrate (vit C) before it reaches the electrode surface. A comparison was performed with the sensor without the enzyme, and the difference in response was used to calculate the vit C selectivity index, thereby underlining vit C detection in a mixture of phenols. Very often, vitamin C is also used as a standard expression of the AC of other analysed AOx [33]. Biosensors for the detection of antioxidant enzymes are less encountered in literature. These enzymes are usually incorporated in the biosensor architecture in order to detect AOx compounds. One such enzyme is peroxidase extracted from a plant known as ice-cream-bean, which is used for the detection of TBHQ [25].

3. Biosensors for the Assessment of Phenol Classes

Most biosensors reported in the literature focus on the detection of TAC, where this review focuses on biosensors or even sensors that discriminate among antioxidant compounds or classes of antioxidant compounds. Of great interest is the detection of total phenolic content (TPC), but also the singular or simultaneous detection of certain antioxidants. We present the recent advances achieved in biosensor technologies for AOx evaluation, directed toward phenolic compounds, and the analytical parameters of each biosensor architecture and their corresponding performances are presented in Table 1.

3.1. Biosensors for Phenolic Content Assessment

The biosensors for phenolic compounds mainly incorporate two enzymes: tyrosinase (Tyr) and/or laccase (Lac). A laccase crude extract from *Pycnoporus sanguineus* was mixed with graphite and mineral oil to obtain a laccase-based modified carbon paste biosensor [34]. The biosensor was able to detect the TPC in red fruit extracts. In another configuration, the carbon paste was modified with organofunctionalized silica and the new biosensor was used to estimate the TPC of honey samples through the enzymatic oxidation of phenolic compounds, which after reaching a plateau, were electrochemically reduced at the electrode surface [35]. The corresponding cathodic peak currents were used to express the TPC value in gallic acid equivalent (μg (GAE).g^{-1}), which were close to the values obtained using the classical assay of Folin-Ciocalteu (FC).

Another simple biosensor architecture was reported by Rodríguez-Sevilla et al. [36]. They immobilized mushroom tyrosinase (Tyr) onto screen-printed electrodes (SPE) using three different techniques: entrapment with water-soluble polyvinyl alcohol (PVA), and crosslinking with glutaraldehyde (GA) in the absence and presence of human serum albumin (HSA). All biosensor configurations were tested in the presence of catechol (CAT), and the best performances were obtained for SPE/Tyr/GA with a sensitivity (S) of 26 ± 4 nA μM^{-1}. Finally, this biosensor was used for the quantification of the Trolox Equivalent Antioxidant Capacity (TEAC) of real medicinal plant samples.

Another simple biosensor design was done by García-Guzmán et al. [37] by modifying a Sonogel-Carbon electrode with Lac from *Trametes versicolor* mixed with GA and Nafion, drop-casting the obtained solution onto the electrode. Phenolic compounds in wines were analysed, first for the detection of individual phenolic compounds (gallic acid; quercetin; rutin; tannic acid; ferulic acid; (+)catechin; (-)epicatechin (ECAT); tyrosol; caffeic acid (CA); vanillic acid; syringic acid; p-coumaric acid and 4-methyl catechol) and second, the total polyphenols content. From the first assay, the authors found that not all selected polyphenols gave an amperometric response; whereas they found good sensitivities for most of the o-diphenols. For the second assay, an increase in the signal compared with the signal obtained for individual phenols was explained either by the synergic effect among the polyphenols or by the contribution from several not individually detectable polyphenols. The same group improved the biosensor a few years later by adding a polymer film, poly(3,4-ethylenedioxythiophene (PEDOT), into the biosensor architecture and using Lac instead of Tyr [38]. Another simple approach based on a carbon paste electrode on which Tyr is immobilized in a Nafion film was presented by Sýs et al. [39]. A mushroom Tyr was used to catalyze the oxidation of p-hydroquinone (HQ) and a satisfactory LoD of 1.6 µM was obtained. An interference study with vitamin C concluded that the vitC:HQ molar concentration ratio should not exceed 1 in order to avoid an interference. For substrate specificity experiments, catechol, resorcinol, phenol and Trolox were used, and the authors concluded that Tyr catalyzes predominantly the oxidation of polyphenols having their hydroxyl group in ortho position. Compounds with this group in meta or para positions need longer time of oxidation, thereby allowing the biosensor to be used for TEAC determinations in wine samples. Another simple and interesting concept was based on α-cyclodextrin modified CPE electrochemical biosensor to detect trans-resveratrol in grape with a sensitivity of S = 310.78 nA µg/L and a limit of detection of LoD = 12 µg/L [40]. The cyclodextrins act as molecular receptors for resveratrol due to their stable "host–guest inclusions" with phenolic substances.

Recent advances in nanomaterial technology have been employed for a variety of biosensors incorporating carbon-based nanomaterials, metallic nanoparticles, or polymers. A solid-contact potentiometric biosensor with two-layer transducer was reported by [41]. The first layer contains a blend of poly(vinyl) chloride carboxylate (PVC-COOH), graphite and potassium permanganate, and the second one contains a mixture of PVC-COOH and graphite, which was deposited on top of the first one. On the last layer, tyrosinase enzyme, extracted from green yellow banana peel, was immobilized by reaction with N-(3-dimethylaminopropyl)-N′-ethylcarbodiimide hydrochloride. Catechol was used as reference phenol, and its concentration was plotted against a relative potential, obtained through the difference between initial and final potential. In the same manner, honey and propolis samples were analysed for their phenolic content, and the results were expressed in mg of GAE by 100 g of sample. For most samples, they found very similar values when compared to the classical Folin–Ciocalteu method. Additionally, using catechol as reference, Cabaj et al. immobilized Tyr in an electrochemically synthesized copolymer based on N-nonylcarbazole derivatives [42]. A platinum electrode was coated with a thin polymer film through electropolymerization, while fungal Tyr from *Agaricus Bisporus* was electrodeposited. A temperature study for Tyr electrodeposition was conducted, and the authors found that T = 47 °C is optimal. They recorded the current response in PBS buffer (pH 7.0) containing 0.2 mM catechol for the temperature range (0–50 °C), and they assigned this high thermal stability to the porosity of the electrosynthesized polymer 2,7-BSeC (poly [2,7-bis(selenophene)-N-nonylcarbazole]), which reduces the conformational flexibility and prevents enzyme unfolding. Both cathechol and L-Dopa were analysed as phenolic compounds, and the sensitivity for catechol was determined to be 2.45 µA mM-1, where LoD was found to be quite low (0.02 µM) compared to other biosensor configurations at that time.

The Langmuir–Blodgett technique was used to develop films containing mushroom tyrosinase, lutetium bisphthalocyanine and arachidic acid (Tyr/AA/LuPc2) adsorbed onto ITO glass [43] and was used for the detection of several phenol derivatives. The main advantage was that the Langmuir–Blodgett film could be cycled up to 50 times for which the percentage of decay was lower than 4%.

Nanoparticles are widely used to either improve the enzymatic catalytic activity, enhance biosensor selectivity or gain more control over the electrode microenvironment [13,16,44]. Carbon-based nanomaterials such as single- or multi-walled carbon nanotubes (SW-, MW-CNT), carbon nanofibers are available on a large scale and are most often incorporated into biosensor architectures. De Oliveira et al. [25] used MWCNT mixed with graphite powder and sepiolite clay mineral containing peroxidase. The homogeneous paste, incorporated into a plastic syringe, was used as a working electrode for the detection of the synthetic TBHQ. Square wave voltammetry (SWV) was used to monitor the reduction potential of TBHQ at the biosensor surface. The LoD and quantification limit (LoQ) were determined to be 0.41 and 1.25 mg L^{-1}. Furthermore, the biosensor was used to determine TBHQ recovery from salad dressing samples, and the results were comparable with those obtained with the classical HPLC technique. Also, the biosensor was sufficiently sensitive to be used in the quality control of foods. MWCNTs alongside reduced graphene oxide (GO) were used by Vlamidis et al. [45] for the detection of polyphenols in fruit juices. Several biosensor architectures were tested and compared, using both laccase from *Trametes versicolor* and tyrosinase from mushroom, immobilized onto GCE, employing different agents such as bovine serum albumin (BSA) and glutaraldehyde as crosslinking agents, chitosan, and Nafion. The Lac-based biosensor showed a better performance towards catechol detection and the biosensor configuration was optimized as follows: GCE was modified with a mixture of reduced GO and MWCNT (denoted as "hybrid" layer), followed by Lac immobilization. To extend the lifetime of the biosensor, several protective biomembranes were tested, and a mixture of 20 mg mL^{-1} BSA and 2.5% (v/v) GA, denoted BSA-GA1 was found to be optimal. Finally, the chronoamperometric response of the biosensor was recorded at 0 V and the calibration graphs were constructed by plotting the reduction current as a function of catechol concentration. This is the first work using Lac for catechol detection in this review, obtaining a LoD of 0.3 μM, while in the case of Tyr, the LoD is slightly higher (0.5 μM). Only the biosensor Pt/2,7-BSeC/Tyr [42] managed to obtain a lower detection limit for catechol using Tyr. The GCE/hybrid/Lac/BSA-GA1 biosensors performance was tested towards a variety of polyphenols (gallic acid, pyrogallol, 2,3-di-hydroxybenzoic acid, dopamine, epicatechin, catechin, rutin, caffeic acid and chlorogenic acid) displaying a good sensitivity towards most compounds. For real sample analysis, fruit juices were analysed comparing the results obtained with Lac- and Tyr-based biosensors for the total polyphenolic content.

The group of Zgardzińska [46] used microporous carbon fibers as electro-conductive immobilization matrixes for Lac. The authors prepared two different types of carbon fibers (CFs) by cellulose carbonization. The differences were mainly in the specific surface area and their preparation process. Both CFs (CFA and CFB) had delignified cellulose as precursor, but for CFB, a $ZnCl_2$ activator was used, flattening the microfibers. Laccase from *Trametes versicolor* was coupled with CFs-modified graphite-rod electrodes, both showing a very similar response during CVs. The optimal working parameters were determined for ABTS (2,2'- Azino-bis(3-ethylbenzthiazoline-6-sulfonic acid)) and catechol. For both compounds, the amperometric detection is based on the reduction of oxidized electroactive products (ABTS cation radicals and semiquinone intermediates) formed during the enzymatic reaction. The detection limit to catechol for both biosensors was less than 5 μM, where the sensitivity for CFA was 1000 A M^{-1} m^{-2} and for CFB 1137 A M^{-1} m^{-2}. The CFA-modified biosensor was used for analysis of catechol in raw and spiked (0.25 mM CAT) wastewater samples and the results were compared with those obtained with HPLC. Between the two techniques, a significant difference could be noticed and was attributed to the fact that wastewater could have contained some catechol-like components that were also oxidized by the enzyme, due to the broad specificity of laccase. When using the "standard addition mode" on the CFA-modified biosensor, its sensitivity towards raw wastewater was very similar to that of catechol standard solution (5.54 ± 0.29 mA mM^{-1} and 5.6 ± 0.4 mA mM^{-1}), demonstrating the high accuracy of biosensing.

Gold nanoparticles (AuNPs) are also widely used due to their enhanced catalytic properties. A layer-by-layer (LbL) biosensor architecture containing AuNPs was developed by Salvo-Comino et al. [47]. Using an ITO electrode, successive immersions of the substrate into electrocatalytic solutions, containing chitosan (CHI), positively charged, and two negatively charged solutions with copper(II) phthalo-cyanine-tetrasulfonic acid tetrasodium salt (CuPcS) and AuNPs were performed. Two LbL architectures were tested, [(CHI)-(AuNP)-(CHI)-(CuPcS)]$_2$ and [(CHI)-(CuPcS)-(CHI)-(AuNP)]$_2$, and on each multilayer covered electrode, either Tyr (from *Agaricus bisporus*) or Lac (from *Trametes versicolor*) was immobilized. The LbL structure with higher roughness and pore size (meaning the presence of (CuPcS) as top layer) was shown to facilitate the diffusion of catechol, and among the two enzymes, Tyr showed the best catalytic properties, obtaining the lowest LoD to date, at 0.85 nM. The use of two enzymes was to better evaluate the electron mediator capability of the LbL, knowing that Tyr oxidizes monophenols and o-diphenols to the corresponding quinone (in a two-electron process), whereas Lac catalyzes the one-electron oxidation of several aromatic substrates. Chronoamperograms were recorded for increasing concentrations of catechol, and a sensitivity of 0.681 A M^{-1} was calculated for the best performing LbL/Tyr configuration. No real sample analysis was done in this work. Another biosensor configuration containing AuNPs alongside graphene nanoplatelets was developed by Zrinski et al. [48]. Graphene nanoplatelet-modified screen-printed carbon electrodes were fabricated by printing with a graphene nanoplatelets modified carbon ink on a ceramic substrate. Next, a solution of AuNPs was drop-casted on the surface, followed by the enzyme solution containing a mixture of nafion, ethanol, water and Lac from *Rhus vernicifera*. After drying, the electrode was used for HQ detection. The electrochemical behaviour of HQ was also evaluated at each modification step, and in the end, the TEAC assay was chosen as the reference method. Great attention was given to the sensor modification, parameter optimization and characterization. The biosensor analytical performances are presented in Table 1, and for repeatability and reproducibility, an RSD (relative standard deviation) of ± 2% and ± 3% was obtained. The authors also made a comparison of their biosensor performances with literature, and where lower LoDs were found, the linear range was shorter. The wider linear range reported in this work was considered suitable for easy evaluation of TEAC in real samples. First, several representative phenolic antioxidants and related compounds were analysed using hydrodynamic amperometry, and the results were presented in a histogram as equimolar ratio of HQ. There was low or no interference with paracetamol, dopamine, ascorbic acid, where a ratio up to 41.2% was found for phenols such as CA, phenol, p-coumaric acid or syringic acid. Using the same method, the total phenolic antioxidant capacity was assessed as Trolox equivalent and hydroquinone equivalent, and the results were compared with those obtained from the TEAC assay. The values obtained with the newly developed Lac biosensor were comparable with those from the conventional spectrophotometric method.

Due to the increased catalytic properties of gold, Liu et al. [49] developed aloe-like Au–ZnO micro/nanoarrays for the detection of catechol. A lot of effort was put into the growth of the aloe-like Au–ZnO arrays on an ITO electrode, and this process is illustrated in Figure 2. Zinc oxide (ZnO) was chosen since it is a well-known semiconducting material on which enzymes can easily adsorb. ZnO/ITO was prepared through a modified wet chemical process, after which the aloe-like ZnO was coated with adsorbed Zn^{2+}; followed by the electrodeposition of gold from a HAuCl$_4$ electrolyte solution.

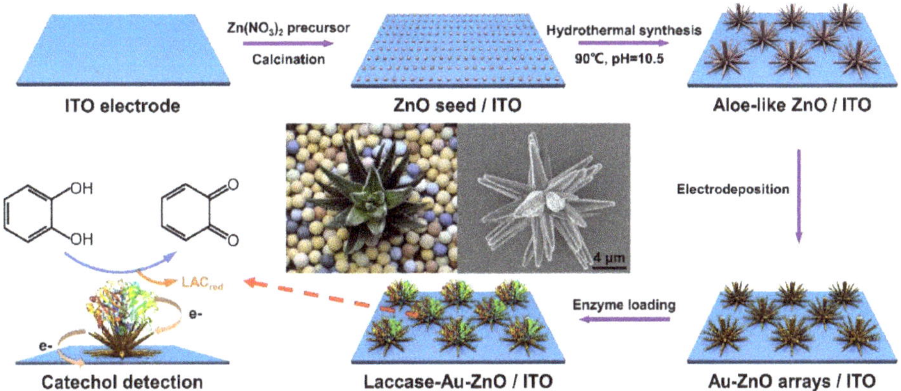

Figure 2. Schematic illustration of the preparation of aloe-like Au–ZnO arrays on an ITO electrode (reproduced from [49] with permission of Elsevier).

Due to electrostatic force, AuCl^{4-} can easily accumulate on the surface of the ZnO. The nanostructure evolution of the arrays was continuously optimized, and the growth mechanism of the crystals was closely monitored in order to obtain the best configuration. The electrochemical behaviour of the Au-ZnO-based electrodes was monitored using electrochemical impedance spectroscopy (EIS), which showed that the introduction of AuNPs efficiently improved the conductivity of ZnO, benefiting of a current signal amplification. After this step, the enzyme was immobilized, and the biosensor performances were tested in the presence of catechol. Since phenol, HQ and resorcinol (RS) may produce interference, the authors tested the influence of these phenolic compounds on the oxidation of catechol. Using CV, only the addition of CAT generated a clearly defined redox signal when compared to the other compounds. Using amperometry in different pH electrolytes, both HQ and CAT generated a current response from the oxidation process at pH 8.25. This happened due to the over-oxidation behaviour of the ZnO material in alkaline environments. However, for the neutral pH of 7.0, only the current change caused by CAT had a significant intensity, while phenol, HQ and RS produced little interference. Herewith, for increasing CAT concentrations, a sensitivity of 131 μA mM^{-1} and a LoD of 25 nM were calculated. The biosensor showed good reproducibility with an RSD of 3.29%, and showed an exceptional stability in the first 10 days, retaining most of its initial sensitivity, starting to decrease up to 15% after 30 days. The biosensor was also tested in real water samples and the results were compared with those of HPLC. No CAT was detected in the water samples, so manual addition was required, and the recovery varied between 98.90% and 101.10%. All results prove that a very sensitive, selective biosensor was developed, and its use in real samples is reliable.

A very interesting and less encountered nanoparticle type was used and functionalized by Palomar et al. [50]. The authors used tungsten disulfide nanotubes (WS$_2$) functionalized with carboxylic acid functions (WS$_2$-COOH). The carboxylic acid groups served as an anchor for the immobilization of Tyr from mushroom via a standard EDC/NHS coupling reaction. The electrochemical behaviour of the WS$_2$-COOH films on GCEs was monitored using CV in organic media, where the irreversible oxidation of WS$_2$ is shown. The WS$_2$-COOH nanotube films with a controlled thickness of 6.2 μm, were then used as a support for the immobilization of tyrosinase, as shown in Figure 3.

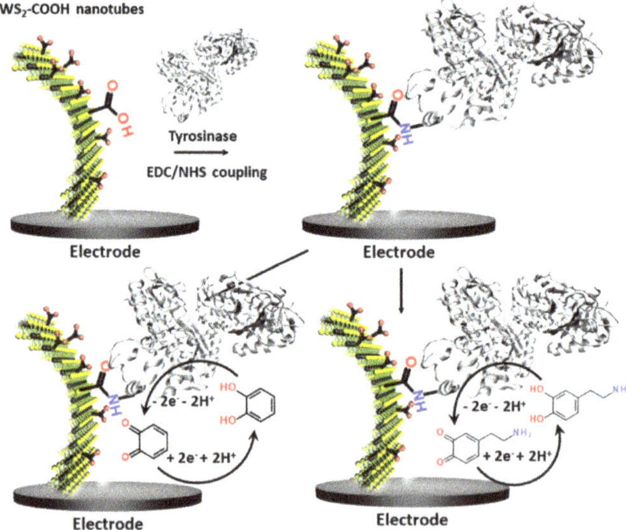

Figure 3. Sketch of the functionalization of WS$_2$ modified glassy carbon electrodes with the enzyme tyrosinase via a standard EDC/NHS coupling reaction. These modified bioelectrodes served in the detection of catechol (bottom left) and dopamine (bottom right) at −0.2 V vs. Ag$^+$/Ag. (reproduced from [50] with permission of RSC Publishing).

For this, the electrode was first incubated with PBS (phosphate buffered saline), pH 7.4, containing EDC (N-(3-Dimethylaminopropyl)-N0-ethylcarbodiimide hydrochloride), NHS (N-hydroxysuccinimide) and DMAP (4-(dimethylamino)pyridine) for 12 h, followed by another immersion of 12 h in Tyr solution. The authors used the newly developed biosensor for the detection of catechol and dopamine due to the enzymatic oxidation of the phenol ring, present in both compounds. Chronoamperometric measurements were done at −0.2 V vs. SCE, and the current density was proportional to the CAT concentration. CAT will be regenerated during the reduction at the electrode, allowing the amplification of the signal, leading to an increased sensitivity of 152.5 mA L cm^{-2} mol^{-1}. Although the sensitivity is quite low compared to other reports, the linear range is similar. This can be explained by the presence of the functionalized WS$_2$ nanotubes, which allow a high amount of enzyme immobilization. The low sensitivity on the other hand was explained either due to rapid oxygen consumption and/or reduced permeability of the WS$_2$-COOH film, as well as lack of conductivity at the used potential. In that same manner as CAT, dopamine was also detected with a sensitivity of 6.2 mA L cm^{-2} mol^{-1} over a linear range between 0.5 and 10 μM. The authors did not obtain spectacular results, but they showed that the WS$_2$-COOH nanotubes can be promising components for the development of electrochemical biosensors.

3.2. DNA- and Protein-Based Biosensors for Phenolic Content Assessment

DNA-based biosensors use DNA as a recognition element, and their performance is assessed based on the principle of oxidative damage inflicted on the DNA by ROS/RNS. By adding an antioxidant to the solution, the oxidative damage should decrease, indirectly evaluating the antioxidant capacity of the analysed phenolic compounds [51]. Either single- or double-stranded DNA can be immobilized on the electrode surface, or one of the amino acids can be used. Wang et al. [52] chose guanine to be added in a composite membrane, as shown in Figure 4.

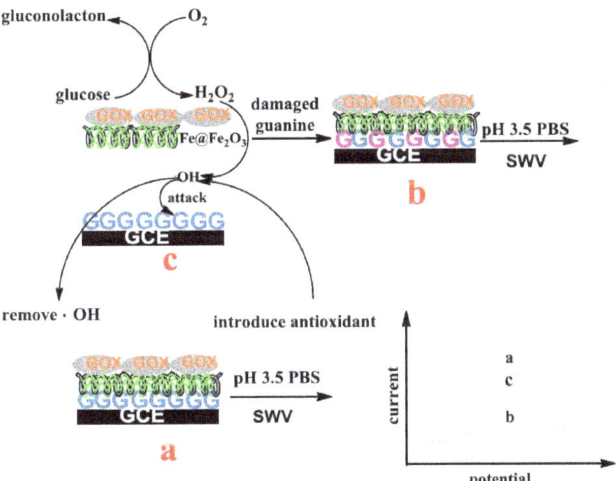

Figure 4. Mechanism of the biosensor and the detection method. SWV (square wave voltammetry). Reproduced from [52] with permission from ESG publisher under a Creative Commons Attribution 4.0 International License http://creativecommons.org/licenses/by/4.0/).

The membrane was built on a GCE by first immersing the electrode into a buffered guanine solution, which adsorbed on the activated GCE surface at a potential of +0.4 V for 180 s under constant stirring. Then, the guanine-modified surface was sequentially covered with Fe@Fe$_2$O$_3$ and glucose oxidase (GOX). The resulting biosensor was named GOX/PDDA-Fe@Fe$_2$O$_3$/G/GCE. The biosensor was built in such way that the reactions between GOX, glucose and Fe@Fe$_2$O$_3$ generated a hydroxyl radical, which oxidized the guanine. SWV was used to determine the biosensor performances, first in PBS, pH 3.5, recording the peak current as i$_0$; and second, in PBS, pH 5.0, containing 50 mmol L^{-1} glucose in the absence or presence of an antioxidant, after incubation for a certain time. The peak current was recorded as i_t, and the guanine oxidative damage, as well as AC, was detected by calculating the signal change for G. Vit C was used as a well-known radical scavenger in order to show the effect of incubation time on the intensity of the peak potential Δi_p (in the presence and absence of vit C). In the presence of vit C, Δi_p was smaller, indicating some protective effects on the oxidation of G. In the same manner, several phenolic compounds were tested, and their AC was expressed as AOT% = $i_t/i_0 \times 100$. The AOT% was determined for CA, coumaric acid and resveratrol, where the highest value was obtained for resveratrol and the lowest for CA. The authors highlight the fact that the G damage process and the protection of phenolic compounds was achieved via a series of biochemical reactions in the composite membrane, simulating the in vivo processes.

Although most DNA-based biosensors found in the literature focus on DNA damage and the detection of TAC, we found two examples where DNA was used as a biopolymer. The first was the work of Mello et al. [53], which describes the use of a DNA additive in the architecture of an enzymatic biosensor, and the focus is on the relationship between total antioxidant activity (TAA) and total phenol content of *Ilex paraguariensis* extracts. This is the only work we found that does not monitor the oxidative damage on DNA, but works as a classical electrochemical biosensor for the detection of chlorogenic acid (CGA). The CPE was prepared by immobilizing the DNA additive and horseradish peroxidase solution (HRP) on silica-titanium with GA. After drying, graphite powder and mineral oil were added to the mixture in order to obtain a homogeneous paste, which was put into a glass tube. The biosensor response was measured as the difference between total and residual current using amperometry. The development and optimization parameters of the biosensor were described in a previous work [54]. The TAA was evaluated by the DPPH colorimetric UV–Vis assay, on the basis of IC_{50}, which represents the AOx concentration needed to reduce 50% of the initial amount of DPPH·. A low IC_{50} value indicates the presence of strong AOx compound in the extracts. By plotting the results obtained for TAA against polyphenols compound concentration using the biosensor method, a linear relationship was observed with correlation coefficients > 0.9 for all analysed samples. Thereby, the authors highlight that the determination of the TPC was representative in terms of AOx compounds, and interferents like ascorbate or carbohydrates showed no response; thus, the phenol content is the main source of AOx in *Ilex paraguariensis* extracts. Another work using DNA as a biopolymer was authored by Ferreira Garcia et al. [55], where the effect of different biosensor modifiers was evaluated on a laccase CPE. The laccase crude extract was obtained through submerged fermentation of *Pycnoporus sanguineus* in a specific growth medium. First, the Lac crude extract was mixed with graphite powder (CPL), to which various modifiers (GA, BSA, chitosan, DNA, silica, titanium dioxide, activated and non-activated CNTs) were added, left to dry, and finally mixed with mineral oil. The homogeneous paste was then filled into an electrode support. After the optimization of parameters and the voltametric characterization of all electrode configurations, the best results were obtained by CPL-DNA:CNT. The presence of activated CNTs significantly improved the biosensor sensitivity, which was further improved by the presence of DNA, whose biocompatibility lead to a better enzymatic activity of the immobilized enzyme. A calibration plot for the detection of rutin was constructed, obtaining an LoD of 12 µM and an LoQ of 38 µM. To further optimize the CPL-DNA:CNT biosensor for the detection of phenolic compounds, the influence of conditioning time and starting potential were evaluated and determined to be optimal at a conditioning time of 30 s and starting potential of 0.5 V. These parameters were kept for an affinity assay; where the relative response of CPL-DNA:CNT against equimolar concentrations of different phenolic compounds (CAT, gallic acid, rutin, CA, CGA, phenol and chlorophenol) was monitored. The highest response was obtained for CAT, closely followed by gallic acid and rutin. For real sample analysis, the total phenolic content in crude coffee samples was determined and compared with the classical Folin–Ciocalteu assay. The total phenolic values for both methods were expressed as GAE mg mL^{-1}, and they were found to be in good agreement, with an RSD% of the biosensor lower than the one obtained for the FC method.

Table 1. Partial applications of various biosensor architectures for the detection of phenolic antioxidants.

Biosensors	Index	Sample	Detection Technique	Linear Range	LoD	Ref
Lac-CPE [1]	TPC	red fruits	DPV	-	-	[34]
Lac-CPE	TPC	honey	DPV	-	-	[35]
SPE [2]/Tyr/GA [3]	catechol	medicinal plants	amperometry	$0 \leq [CAT] \leq 136$ μM	1.5 ± 0.6 μM	[36]
Lac/SNGC [4]	gallic acid	wine	amperometry	-	0.011 mg L^{-1}	[37]
PEDOT [5]-Tyr/SNGC	caffeic acid	beer, wine	amperometry	$10 \leq [CA] \leq 300$ μM	4.33 μM	[38]
CPE/Tyr/Nafion	hydroquinone	red wine	amperometry	$20 \leq [HQ] \leq 120$ μM	1.6 μM	[39]
[6] α-CD-CPE	trans-resveratrol	grape extracts	DPV	$30 \leq [resv] \leq 1000$ μg L^{-1}	12 μg L^{-1}	[40]
Transducer -Tyr	catechol	honey, propolis	potentiometry	$9.3 \times 10^{-7} \leq [CAT] \leq 8.3 \times 10^{-2}$ M	7.3×10^{-7} M	[41]
Pt [7]/2,7-BSeC [8]/Tyr	catechol	-	DPV	$1.5 \leq [CAT] \leq 80$ μM	0.02 μM	[42]
Tyr/AA [9]/LuPc$_2$ [10]	caffeic acid	-	CV	$10 \leq [CA] \leq 400$ μM	1.98 μM	[43]
CPE–CNT [11]–SEP [12]–nafion–peroxidase	TBHQ	salad dressing samples	SWV	$1.65 \leq [TBHQ] \leq 9.82$ mg L^{-1}	0.41 mg L^{-1}	[25]
GCE [13]/hybrid/Lac/BSA [14]-GA1	catechol	fruit juices	amperometry	$1 \leq [CAT] \leq 300$ μM	0.3 μM	[45]
CFA [15]-CFB [16]/Lac	catechol	wastewater	amperometry	-	< 5 μM	[46]
[(CHI [17])-(AuNP [18])-(CHI)-(CuPcS [19])]$_2$-Tyr	catechol	-	amperometry	$2.4 \leq [CAT] \leq 20$ μM	8.55×10^{-4} μM	[47]
LACC/AuNP/GNPl [20]/SPCE [21]	hydroquinone	wine, blueberry syrup	hydrodynamic amperometry	$4 \leq [HQ] \leq 130$ μM	1.5 μM	[48]
Lac/Au–ZnO [22]/ITO [23]	catechol	environmental water	amperometry	$0.075 \leq [CAT] \leq 1100$ μM	25 nM	[49]
GC [24]/WS$_2$-COOH [25]/tyrosinase	catechol	-	amperometry	$0.6 \leq [CAT] \leq 70$ μM	-	[50]
CPL [26]-DNA [27]:CNT	rutin	coffee	DPV	-	12 μM	[55]
protein-based solid biosensor with Cu(II)-Nc [28] assay	epicatechin	herbal infusions	absorbance	$12.5 \leq [ECAT] \leq 150$ μM	1.2 μM	[56]
protein-based solid biosensor with Fe(II)-Fz [29] assay	epicatechin	herbal infusions	absorbance	$25 \leq [ECAT] \leq 250$ μM	0.5 μM	[57]

[1] Carbon paste electrode; [2] Screen-printed electrode; [3] Glutaraldehyde; [4] Sonogel carbon electrode; [5] Poly(3,4-ethylenedioxythiophene; [6] α-cyclodextrin; [7] Platinum; [8] Poly[2,7-bis(selenophene)-N-nonylcarbazole]; [9] Arachidic acid; [10] Lutetium bisphtalocyanine; [11] Carbon nanotubes; [12] Sepiolite; [13] Glassy carbon electrode; [14] Bovine serum albumin; [15] Carbon fibre A; [16] Carbon fibre B; [17] Chitosan; [18] Gold nanoparticle; [19] Copper(II) phthalo-cyanine-tetrasulfonic acid tetrasodium salt; [20] Graphene nanoplatelets; [21] Screen printed carbon electrode; [22] Gold-zinc oxide; [23] Indium tin oxide; [24] Glassy carbon; [25] Tungsten disulphide nanotubes with carboxylic acid functions; [26] Carbon paste laccase; [27] Deoxyribonucleic acid; [28] Copper(II)- neocuproine; [29] Iron(II)-ferrozine.

Returning to the principle of oxidative damage detection, protein-based sensors are an alternative to DNA-based biosensors. In the group of R. Apak, Akyüz et al. published two papers on protein-based solid biosensors, evaluating protein damage. The first biosensor was used for the determination of Cu(II)-induced pro-oxidant activity of several phenolic compounds alongside non-phenolic vitamin C [56]. The protein-based solid sensor was prepared by completely separating the egg white from the yolk, to which water and $CaCl_2$ was added dropwise as protein precipitation agent. The Ca−proteinate precipitate was filtered, washed, dried and grinded, after which the dried protein residue was ready to use. The powder was diluted in phosphate buffer, pH 7.4, copper(II) solution and various concentrations of standard antioxidant solutions or herbal plant infusions. The final solution was vortexed, centrifuged, and the upper liquid decanted. The remaining protein-based sensor was washed and solutions of neocuproine (Nc), NH_4Ac buffer and water were added. The solution was again agitated, filtered and the absorbance at 450 nm was recorded against a reagent blank. The blank was missing the antioxidant standard. Such an assay involves the reduction of Cu(II) ions to Cu(I) due to the presence of the antioxidant compounds, where Cu(I) binds to the solid biosensor. The protein-bound Cu(I), is an indicator of the pro-oxidant activity of antioxidants on proteins, which can be colorimetrically determined through absorbance measurements at 450 nm with Nc. Phenolic compounds from several classes were chosen for analysis: phenolic acids (gallic acid and CGA), flavanols (CAT and ECAT) and flavonols (quercetin—QUE and myricetin—MYR). The pro-oxidant activity of these compounds was generally visible above the critical concentration of 2.50 µM. ECAT was chosen as a standard compound, and analytical performances such as LoD (1.2 µM) and LoQ (4.0 µM) were determined using two methods, the biosensor and the carbonyl assay (considered a standard method as comparison for pro-oxidative status detection), where the biosensor assay was shown to be more sensitive. Vit C had the least pro-oxidant activity, while CAT and ECAT had similar effects and QUE showed the highest activity. For statistical comparison of the two methods, the total pro-oxidant activities of ECAT standard and sage extract were calculated as ECAT equivalent. The obtained pro-oxidant activities can be easily converted to antioxidant activity (given as the ratio between the slopes of the calibration plot for the test compound and reference compound); however, these values were not calculated by the authors. The recovery values were close to 100%, demonstrating that the solid biosensor used with the Cu(II)-Nc assay was suitable to determine the pro-oxidant activity of plant extracts containing polyphenolic compounds. The authors also highlight a close relationship between the pro-oxidative behaviour and the ion reducing ability of transition metals of phenolic compounds. In their second work [57], the group uses the same solid biosensor for developing a novel Fe(III)-ferrozine (Fz) spectrophotometric pro-oxidant activity assay. The Fe(III) reducing ability generated by AOx compounds is an indirect measure of their pro-oxidant activity, since reduced ions may generate ROS. The formed Fe(II) will bind to the solid biosensor, enabling the measurement of the absorbance at 562 nm of the Fe(II)-ferrozine chelate. The solutions were prepared in the same manner, only the buffer was adjusted to pH 5.5, and Nc was replaced with Fz. Again, ECAT was chosen as the standard compound and analytical performances such as LoD (0.5 µM) and LoQ (1.8 µM) were determined. Compared to previous Cu(II)-Nc assay, the performances were improved with ~58%. The pro-oxidant activity of several phenolic compounds was determined, followed by ECAT recoveries from herbal infusions. Although a very elaborated assay as compared to what a biosensor stands for, quite low values for the detection limits of ECAT were obtained.

3.3. Biosensor Trends and Perspectives

This chapter refers to new concepts, trends and promising perspectives, where a classical "biosensor" will not always be described. In response to the demand for fast, sensitive and selective techniques, bioelectronic tongues (BioETs) have been designed, combining achievements of chemometric analysis with the unique properties of biological compounds. Such a device is described by Medina-Plaza et al. [58], based on a previously presented work [43]. Based on the results obtained using principal component analysis (PCA), which demonstrated that a multisensor system was able to discriminate phenols according to the number of phenolic groups present in the structure, the authors formed an array of three electrodes in order to discriminate among phenolic antioxidants in the food industry. The three (bio)sensors were developed as follows: Langmuir–Blodgett (LB) films were prepared using a PBS-NaCl subphase, onto which a mixture of arachidic acid and lutetium bisphtalocyanine was spread. Twenty such monolayers were deposited onto an ITO glass surface, obtaining the first sensor (AA/LuPc$_2$). For the other two enzymatic biosensors, onto 10 monolayers of AA/LuPc$_2$, 10 monolayers of enzyme/AA/LuPc$_2$ were added. Two enzymes were used, Lac form *Trametes versicolor* and Tyr from mushroom. After preparation, the Enz/AA/LuPc$_2$ films were treated with glutaraldehyde to covalently immobilize the enzyme. Cyclic voltammetry was used to evaluate the response of the three-electrode array towards six phenolic compounds: a monophenol (vanillic acid-VA), two ortho-diphenols (CAT and CA), one para-diphenol (HQ) and two triphenols (gallic acid and pyrogallol). The responses, calculated from the peaks associated with AOx, were highly reproducible, lower than 1.5%. The reproducibility of different sensors containing the same enzyme was also found to be lower than 1.75%. The AA/LuPc$_2$ film showed a quasi-reversible and intense redox pair, associated with the presence of LuPc$_2$, which acts as a mediator, amplifying the signal. It can be also noticed, that the LuPc$_2$ redox pair appears at different positions with varying intensities, specific to each analysed compound. The authors also found that the presence of the enzyme increased even more the peak intensities associated with the phtalocyanine ring (~ -0.2 V). From the presented voltammograms, they also concluded that they attained an important degree of cross-selectivity due to the array of sensors. The authors observed significant differences in the response of the Enz/AA/LuPc$_2$ electrodes, attributed to the enzyme specificity, on which a nice discussion was elaborated. The detection limits were determined for each biosensor by measuring the associated peak intensities towards increasing concentrations of the phenolic compounds. The peak intensity linearly increased with the AOx concentration, and the lowest LoD was calculated for gallic acid with the Lac/AA/LuPc$_2$ biosensor, with a value of 4.1×10^{-8} mol L^{-1}. Overall, the detection limits for Enz/AA/LuPc$_2$ were at least one order of magnitude lower than for AA/LuPc$_2$. In the case of CAT, for both Enz/AA/LuPc$_2$ biosensors the LoDs were calculated to be around $\sim 4.7 \times 10^{-7}$ mol L^{-1}, values that are among the lowest (Table 1). Since the array of voltametric electrodes generated signals of intrinsic complexity and cross-selectivity, the authors were able to discriminate the phenolic compounds using PCA. The obtained PCA score plot showed separated clusters (three clusters corresponding to the mono-, di- and tri-phenols). To validate the results obtained with PCA, the BioET was used in musts prepared from grapes of different varieties. The voltammograms clearly showed the response of the phenolic groups present in must, and the peak intensities and positions were related with the total polyphenol index measured by standard chemical methods, which depends on the grape variety. Applying PCA, the biosensor array was able to discriminate grapes according to the grape variety, and the results were validated with the chemical composition of the grape juice.

Another voltametric BioET was described by Ceto et al. [59], comprising an array of four enzyme-modified (bio)sensors. The electrodes were prepared by mixing resin with its corresponding hardener in a ratio of 20:3 (w/w), then adding 15% (w/w) graphite and 2% (w/w) modifier (either Tyr, Lac or copper NPs); after which the mixture was homogenized and left to dry. One electrode was kept as a blank, without any modifier. The electrochemical cell was formed by the four (bio)sensors as a working electrode array, with a double junction Ag/AgCl reference and a platinum counter electrode. The obtained data were further processed with various chemometric tools, including PCA. Since the (bio)sensor array had been already successfully used in the resolution of phenolic compound mixtures in wine samples, the authors proceeded to directly analyse 20 varieties of wine, by recording a complete voltammogram with each sensor for each sample. The working principle of the BioET is presented in a schematic approach in Figure 5.

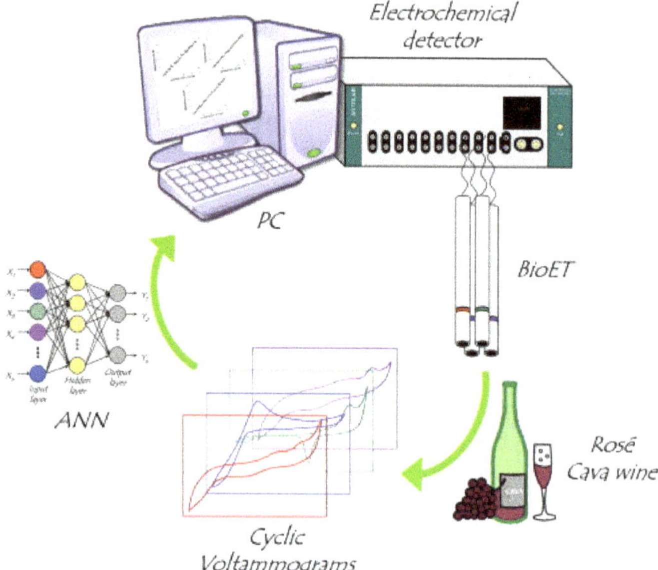

Figure 5. Schematic representation of the BioET approach (reproduced from [59] with permission of Elsevier Publishing).

Overall, the oxidation of the phenolic compounds was noticed in all cases, while the reducing behaviour occurred only in some of the samples. Obtaining a large amount of data with variability among them and among the different biosensors, the condition for developing an ET is fulfilled. Since the BioET array clearly recognizes phenolic content, PCA was used to detect any similarities between the obtained voltametric responses. Thereby, the PCA plot shows how the samples group according to their phenolic content. It could be clearly observed that one cluster groups samples with low phenolic content, while another cluster groups the samples with the highest phenolic content, obtaining six clusters in total. No specification was made on any specific phenolic compound, all the data being regarded as the total phenolic index. To validate the BioET response, modelling software was used (ANN), about which we will only specify that the theoretical expected values were very close to the predicted ones.

An interesting topic worth debating is the replacement of biocomponents with NPs or polymers capable of mimicking the biological component in question. Such systems gain more and more attention due to increased stability and long-term use. Cui et al. [60] found a promising alternative for natural enzymes in a catalytically active nanomaterial by synthesizing a porphyrin-based porous organic polymer, named FePPOP-1. Porphyrin-based porous organic polymers (PPOPs) present a large specific surface area, tuneable pore structures and high stability, properties which facilitate electron and mass transfer. Introducing metalloporphyrins in a POPs framework, the PPOPs skeleton itself will act as a catalyst, integrating the polymeric structure into the family of nanozymes. FePPOP-1 was obtained by the reaction between iron 5,10,15,20-tetrakis-(4'-bromophenyl) porphyrin and 1,3,5-triethynylbenzene, presenting a high peroxidase activity and low LoDs, as will be shown. To verify the peroxidase-like activity of FePPOP-1, the authors used the typical peroxidase substrate of 3,3',5,5'-Tetramethylbenzidine (TMB) in the presence of H_2O_2, where FePPOP-1 catalyses the oxidation of TMB. The decomposition of H_2O_2 into $\cdot OH$ radicals was highlighted by using terephthalic acid as a fluorescent probe. The authors presented the following reaction mechanism: FePPOP-1 catalyses the decomposition of H_2O_2 into $\cdot OH$, which oxidizes the TMB substrate. By varying the concentration of FePPOP-1, the fluorescence intensity gradually increased, proving H_2O_2 decomposition. The detection of H_2O_2 was achieved by adding different concentrations into a mixture of acetate buffer (pH 3.8), TMB and FePPOP-1, after which UV–Vis spectra were recorded. Since the catalytic activity of FePPOP-1 depends on H_2O_2 concentration, a linear dependence of the TMB absorbance with increasing H_2O_2 concentration was observed, and an LoD of 6.5 µM was determined. H_2O_2 is also considered a free radical, which means that in the presence of an antioxidant species, H_2O_2 scavenging should occur. Thereby, the antioxidant capacity of the phenolic compounds gallic acid and tannic acid (TA), alongside vit C was achieved through indirect detection. First, it should be mentioned that in the AOx-free FePPOP-1-TMB mixture, TMB oxidation resulted in a blue colour, which faded upon AOx addition. Thereby, UV–Vis absorption allows the colorimetric detection of AOx. To highlight this process, the AOx were mixed with acetate buffer, TMB and FePPOP-1. Next, a fixed concentration of 50 mM H_2O_2 was added to the mixture for 3 min and the UV–Vis spectra were recorded. The absorbance intensity decreased with increasing AOx concentration. After construction, the calibration plots for each compound, vit C, gallic acid and TA, could be detected to be as low as 0.35, 0.75 and 0.048 µM. The antioxidative abilities of the compounds were evaluated in the following order: TA > vit C > gallic acid. Table 1 presents only one work [37] that uses gallic acid as a standard, and if we compare the two LoDs, this assay is more sensitive compared to the amperometric assay.

The concept of indirect detection of AOx capacity has been widely discussed in the literature, and although it is not the main topic in this review, in addition to the previous reference, we found another interesting biosensor architecture based on the indirect detection of gallic acid via H_2O_2 inhibition. An amperometric biosensor using Prussian Blue (PB) and xanthine oxidase (XOD) was developed by Becker et al. [61]. SPE were first modified with PB, which acts as a suitable mediator for H_2O_2 reduction. Next, a homogeneous mixture containing XOD solution and Azide-unit Pendant Water-soluble Photopolymer (PVA/AWP) was spread on the SPE. The electrode was then exposed to neon light at +4 °C to allow polymerization. Amperometric measurements were performed in a dark glass cell in K-PBS at a working potential of -0.1 V vs. Ag/AgCl. After signal stabilization, 5 mM hypoxanthine (HX—the specific substrate for XOD) was successively added. This procedure was repeated in the absence and presence of different AOx samples, and the corresponding calibration plots were constructed. The evaluation of the AOx capacity is based on monitoring the H_2O_2 produced during the oxidation of HX to uric acid in the presence of the enzyme. The generated H_2O_2 is then reduced at the polarized surface of the biosensor. The addition of AOx reduces the H_2O_2 concentration. The biosensor was used for the determination of gallic acid antioxidant capacity, by recording the amperometric response as a function of successive addition of 5 mM HX in the absence and presence of several gallic acid concentrations ranging between 12.5 and 200 µM. With increasing AOx concentration, the current signal decreased and its AC correspondent was expressed in percentage. For a concentration of 200 µM, an AC of 42% was calculated. The authors presented further analytical data on the biosensor performance for HX detection in the absence and presence of 12.5 µM gallic acid as follows: the biosensor response was linear only in the absence of gallic acid in the range of 10–75 µM HX, the LoDs were 2.17 and 0.72 µM in the absence and presence of gallic acid, and the LoQ also decreased from 7.15 to 2.42 µM in the presence of gallic acid. The biosensor was then successfully applied in real samples of Amazonian fruits.

4. Conclusions

Although a large number of publications focus on the evaluation of TAC, using direct or indirect (ROS scavenging) methods, we managed to bring together several publications focusing on the specific detection of a phenolic compound, with a final chapter dedicated to concepts and methodologies worth mentioning due to their novelty, complexity or improved performance. Most described biosensors were also applied in real sample analysis, mainly related to food or health industries. The majority of biosensors were electrochemical, using mostly amperometric, followed by voltametric (DPV) techniques to evaluate their performances. The newest trends are based on the synthesis of novel materials or combination of various nanomaterials due to their physical, chemical or optical properties.

The authors also look to replace biological compounds with materials which are perfectly capable of mimicking the biological activity (e.g., nanozymes).

Even though the performance of the biosensors is continuously improving, their miniaturization and portability is still an issue that needs to be overcome. Since most of the described biosensors have applicability in the food industry, their portability should be an important aspect.

A rigorous comparison and analysis of all presented biosensors is difficult to obtain due to the many differences and variables determined by each research group. Even though many biosensors use laccase, the enzyme is extracted from various sources, introducing already a factor of uncertainty. Catechol is one of the main phenolic standards used in seven assays, and the LoD varies from 8.55×10^{-4} to 5 μM. A difference of 4 orders of magnitude is quite significant, though different enzymes, NPs, polymers and detection techniques were employed.

Biosensors are therefore a fast, reliable alternative to classical techniques. Depending on the scope of research, however, biosensors can be used as a complementary technique alongside classical ones, especially when the focus of research is based on understanding the role of each molecule in a matrix. Plant matrices are very complex, and as some reviewed articles highlighted, there are scenarios where biosensors detect a class of phenols or TPC, depending on their architecture. On the other hand, focusing on specific analytes, biosensors have the advantage to lower the detection limit and increase selectivity and specificity using various nanomaterials or polymers. The versatility of (bio)sensors is of great importance, since they can be used independent or combined with standard classical methods depending on the scope of research.

Author Contributions: All authors were involved in conceptualization, writing review and editing, supervision, and funding acquisition. All authors have read and agreed to the published version of the manuscript.

Funding: This work was supported by a grant of the Romanian Ministry of Research and Innovation, UEFISCDI-CCCDI, under project number PN-III-P1-1.2-PCCDI-2017-0062, contract no. 58 and grant CNCS - UEFISCDI, project number PN-III-P1-1.1-PD-2019-1285, within PNCDI III.

Conflicts of Interest: The authors declare no conflict of interest.

References

1. Dasgupta, A.; Klein, K. *Antioxidants in Food, Vitamins and Supplements*; Elsevier: Amsterdam, The Netherlands, 2014.
2. Badarinath, A.V.; Rao, K.M.; Madhusudhana, C.C.; Ramkanth, S.; Rajan, T.V.S.; Gnanaprakash, K. A Review on In-vitro Antioxidant Methods: Comparisions, Correlations and Considerations. *Int. J. Pharmtech Res.* **2010**, *2*, 1276–1285.
3. Lobo, V.; Patil, A.; Phatak, A.; Chandra, N. Free radicals, antioxidants and functional foods: Impact on human health. *Pharmacogn. Rev.* **2010**, *4*, 118–126. [CrossRef]
4. Senoner, T.; Dichtl, W. Oxidative Stress in Cardiovascular Diseases: Still a Therapeutic Target? *Nutrients* **2019**, *11*, 2090. [CrossRef] [PubMed]
5. Asmat, U.; Abad, K.; Ismail, K. Diabetes mellitus and oxidative stress-A concise review. *Saudi Pharm. J.* **2015**, *24*, 547–553. [CrossRef] [PubMed]
6. Gilgun-Sherki, Y.; Melamed, E.; Offen, D. Oxidative stress induced-neurodegenerative diseases: The need for antioxidants that penetrate the blood brain barrier. *Neuropharmacology* **2001**, *40*, 959–975. [CrossRef]
7. Pohanka, M. Alzheimer's disease and oxidative stress: A review. *Curr. Med. Chem.* **2014**, *21*, 356–364. [CrossRef] [PubMed]
8. Zuo, T.; Zhu, M.; Xu, W. Roles of Oxidative Stress in Polycystic Ovary Syndrome and Cancers. *Oxidative Med. Cell. Longev.* **2015**, *2016*, 1–14. [CrossRef] [PubMed]
9. Cairns, R.A.; Harris, I.S.; Mak, T.W. Regulation of cancer cell metabolism. *Nat. Rev. Cancer* **2011**, *11*, 85–95. [CrossRef] [PubMed]
10. Nimse, S.; Pal, D. Free radicals, natural antioxidants, and their reaction mechanisms. *RSC Adv.* **2015**, *5*, 27986–28006. [CrossRef]
11. Prior, R.L.; Wu, X.; Schaich, K. Standardized Methods for the Determination of Antioxidant Capacity and Phenolics in Foods and Dietary Supplements. *J. Agric. Food Chem.* **2005**, *53*, 4290–4302. [CrossRef] [PubMed]

12. Buratti, S. A low-cost and low-tech electrochemical flow system for the evaluation of total phenolic content and antioxidant power of tea infusions. *Talanta* **2008**, *75*, 312–316. [CrossRef] [PubMed]

13. David, M.; Şerban, A.; Popa, C.; Florescu, M. A Nanoparticle-Based Label-Free Sensor for Screening the Relative Antioxidant Capacity of Hydrosoluble Plant Extracts. *Sensors* **2019**, *19*, 590. [CrossRef] [PubMed]

14. Suryakumar, G.; Gupta, A. Medicinal and therapeutic potential of Sea buckthorn (Hippophae rhamnoides L.). *J. Ethnopharmacol.* **2011**, *138*, 268–278. [CrossRef] [PubMed]

15. Oroian, M.; Escriche, I. Antioxidants: Characterization, natural sources, extraction and analysis. *Food Res. Int.* **2015**, *74*, 10–36. [CrossRef]

16. David, M.; Serban, A.; Radulescu, C.; Danet, A.F.; Florescu, M. Bioelectrochemical evaluation of plant extracts and gold nanozyme-based sensors for total antioxidant capacity determination. *Bioelectrochemistry* **2019**, *129*, 124–134. [CrossRef]

17. Bunaciu, A.A.; Danet, A.F.; Fleschin, S.; Aboul-Enein, H.Y. Recent Applications for in Vitro Antioxidant Activity Assay. *Crit. Rev. Anal. Chem.* **2015**, *46*, 389–399. [CrossRef]

18. Wantusiak, P.M.; Glód, B.K. Application of UV detection in HPLC in the total antioxidant potential assay. *Open Chem.* **2012**, *10*, 1786–1790. [CrossRef]

19. Zhang, Y.; Li, Q.; Xing, H.; Lu, X.; Zhao, L.; Qu, K.; Bi, K. Evaluation of antioxidant activity of ten compounds in different tea samples by means of an on-line HPLC–DPPH assay. *Food Res. Int.* **2013**, *53*, 847–856. [CrossRef]

20. Maia, L.F.; Ferreira, G.R.; Costa, R.C.C.; De Lucas, N.C.; Teixeira, R.I.; Fleury, B.G.; Edwards, H.G.M.; De Oliveira, L.F.C. Raman Spectroscopic Study of Antioxidant Pigments from Cup Corals *Tubastraea* spp. *J. Phys. Chem. A* **2014**, *118*, 3429–3437. [CrossRef]

21. Jantasee, A.; Thumanu, K.; Muangsan, N.; Leeanansaksiri, W.; Maensiri, D. Fourier Transform Infrared Spectroscopy for Antioxidant Capacity Determination in Colored Glutinous Rice. *Food Anal. Methods* **2013**, *7*, 389–399. [CrossRef]

22. Ye, Y.; Ji, J.; Sun, Z.; Shen, P.; Sun, X. Recent advances in electrochemical biosensors for antioxidant analysis in foodstuff. *TrAC Trends Anal. Chem.* **2020**, *122*, 115718. [CrossRef]

23. Aziz, M.A.; Diab, A.S.; Mohammed, A.A. Antioxidant Categories and Mode of Action. In *Antioxidants*; IntechOpen: London, UK, 2019.

24. Handique, J.; Baruah, J.B. Polyphenolic compounds: An overview. *React. Funct. Polym.* **2002**, *52*, 163–188. [CrossRef]

25. De Oliveira, T.R.; Grawe, G.F.; Moccelini, S.K.; Terezo, A.J.; Castilho, M. Enzymatic biosensors based on ingá-cipó peroxidase immobilised on sepiolite for TBHQ quantification. *Analyst* **2014**, *139*, 2214. [CrossRef]

26. Pandey, K.B.; Rizvi, S.I. Plant polyphenols as dietary antioxidants in human health and disease. *Oxidative Med. Cell. Longev.* **2009**, *2*, 270–278. [CrossRef] [PubMed]

27. Baptista, R.; Madureira, A.M.; Jorge, R.; Adão, R.; Duarte, A.; Duarte, N.; Lopes, M.M.; Teixeira, G. Antioxidant and Antimycotic Activities of Two Native Lavandula Species from Portugal. *Evid. Based Complement. Altern. Med.* **2015**, *2015*, 1–10. [CrossRef] [PubMed]

28. Almajano, M.-P.; Carbo, R.; Jiménez, J.A.L.; Gordon, M. Antioxidant and antimicrobial activities of tea infusions. *Food Chem.* **2008**, *108*, 55–63. [CrossRef]

29. Gil, E.D.S.; Couto, R.O.D. Flavonoid electrochemistry: A review on the electroanalytical applications. *Rev. Bras. de Farm.* **2013**, *23*, 542–558. [CrossRef]

30. Masa, A.; Vilanova, M.; Pomar, F. Varietal differences among the flavonoid profiles of white grape cultivars studied by high-performance liquid chromatography. *J. Chromatogr. A* **2007**, *1164*, 291–297. [CrossRef]

31. Pereira, J.A.; Oliveira, I.; Sousa, A.; Valentão, P.; Andrade, P.B.; Ferreira, I.; Ferreres, F.; Bento, A.A.; Seabra, R.; Estevinho, L.M. Walnut (Juglans regia L.) leaves: Phenolic compounds, antibacterial activity and antioxidant potential of different cultivars. *Food Chem. Toxicol.* **2007**, *45*, 2287–2295. [CrossRef]

32. Torrinha, Á.; Amorim, C.G.; Montenegro, M.C.B.S.M.; Araujo, A.N. Biosensing based on pencil graphite electrodes. *Talanta* **2018**, *190*, 235–247. [CrossRef]

33. Tienaho, J.; Sarjala, T.; Franzén, R.; Karp, M. Method with high-throughput screening potential for antioxidative substances using Escherichia coli biosensor katG′::lux. *J. Microbiol. Methods* **2015**, *118*, 78–80. [CrossRef] [PubMed]

34. De Macêdo, I.Y.L.; Garcia, L.F.; Neto, J.R.D.O.; Leite, K.C.D.S.; Ferreira, V.S.; Ghedini, P.C.; Gil, E.D.S. Electroanalytical tools for antioxidant evaluation of red fruits dry extracts. *Food Chem.* **2017**, *217*, 326–331. [CrossRef] [PubMed]

35. Neto, J.R.D.O.; Rezende, S.G.; Lobón, G.S.; Garcia, T.A.; De Macêdo, I.Y.L.; Garcia, L.F.; Alves, V.F.; Torres, I.M.S.; Santiago, M.F.; Schimidt, F.; et al. Electroanalysis and laccase-based biosensor on the determination of phenolic content and antioxidant power of honey samples. *Food Chem.* **2017**, *237*, 1118–1123. [CrossRef] [PubMed]

36. Rodríguez-Sevilla, E.; Ramírez-Silva, M.-T.; Romero-Romo, M.; Ibarra-Escutia, P.; Palomar-Pardavé, M. Electrochemical Quantification of the Antioxidant Capacity of Medicinal Plants Using Biosensors. *Sensors* **2014**, *14*, 14423–14439. [CrossRef]

37. García-Guzmán, J.J.; Hernández-Artiga, M.P.; De León, L.P.-P.; Bellido-Milla, D. Selective methods for polyphenols and sulphur dioxide determination in wines. *Food Chem.* **2015**, *182*, 47–54. [CrossRef]

38. García-Guzmán, J.J.; López-Iglesias, D.; Cubillana-Aguilera, L.; Lete, C.; Lupu, S.; Palacios-Santander, J.M.; Bellido-Milla, D. Assessment of the Polyphenol Indices and Antioxidant Capacity for Beers and Wines Using a Tyrosinase-Based Biosensor Prepared by Sinusoidal Current Method. *Sensors* **2018**, *19*, 66. [CrossRef]

39. Sýs, M.; Pekec, B.; Kalcher, K.; Vytřas, K. Amperometric Enzyme Carbon Paste-Based Biosensor for Quantification of Hydroquinone and Polyphenolic Antioxidant Capacity. *Int. J. Electrochem. Sci.* **2013**, *8*, 9030–9040.

40. Pekec, B.; Oberreiter, A.; Hauser, S.; Kalcher, K.; Ortner, A. Electrochemical Sensor Based on a Cyclodextrin Modified Carbon Paste Electrode for Trans -Resveratrol Analysis. *Int. J. Electrochem. Sci.* **2012**, *7*, 4089–4098.

41. Draghi, P.F.; Fernandes, J.C.B. Label-free potentiometric biosensor based on solid-contact for determination of total phenols in honey and propolis. *Talanta* **2017**, *164*, 413–417. [CrossRef]

42. Cabaj, J.; Jędrychowska, A.; Swist, A.; Sołoducho, J. Tyrosinase Biosensor for Antioxidants Based on Semiconducting Polymer Support. *Electroanalysis* **2016**, *28*, 1383–1390. [CrossRef]

43. Apetrei, C.; Alessio, P.; Constantino, C.; De Saja, J.; Rodriguez-Mendez, M.L.; Pavinatto, F.; Fernandes, E.G.R.; Zucolotto, V.; Oliveira, O. Biomimetic biosensor based on lipidic layers containing tyrosinase and lutetium bisphthalocyanine for the detection of antioxidants. *Biosens. Bioelectron.* **2011**, *26*, 2513–2519. [CrossRef] [PubMed]

44. Dhand, C.; Dwivedi, N.; Ying, A.N.J.; Lakshminarayanan, R.; Ramakrishna, S.; Loh, X.J.; Verma, N.K.; Beuerman, R.W. Methods and strategies for the synthesis of diverse nanoparticles and their applications: A comprehensive overview. *RSC Adv.* **2015**, *5*, 105003–105037. [CrossRef]

45. Vlamidis, Y.; Gualandi, I.; Tonelli, D. Amperometric biosensors based on reduced GO and MWCNTs composite for polyphenols detection in fruit juices. *J. Electroanal. Chem.* **2017**, *799*, 285–292. [CrossRef]

46. Kavetskyy, T.; Smutok, O.; Demkiv, O.; Maťko, I.; Švajdlenková, H.; Šauša, O.; Novák, I.; Berek, D.; Čechová, K.; Pecz, M.; et al. Microporous carbon fibers as electroconductive immobilization matrixes: Effect of their structure on operational parameters of laccase-based amperometric biosensor. *Mater. Sci. Eng. C* **2020**, *109*, 110570. [CrossRef] [PubMed]

47. Salvo-Comino, C.; González-Gil, A.; Rodriguez-Valentin, J.; García-Hernandez, C.; Martín-Pedrosa, F.; Garcia-Cabezon, C.; Rodriguez-Mendez, M.L. Biosensors Platform Based on Chitosan/AuNPs/Phthalocyanine Composite Films for the Electrochemical Detection of Catechol. The Role of the Surface Structure. *Sensors* **2020**, *20*, 2152. [CrossRef] [PubMed]

48. Zrinski, I.; Pungjunun, K.; Martinez, S.; Zavašnik, J.; Stanković, D.; Kalcher, K.; Mehmeti, E. Evaluation of phenolic antioxidant capacity in beverages based on laccase immobilized on screen-printed carbon electrode modified with graphene nanoplatelets and gold nanoparticles. *Microchem. J.* **2020**, *152*, 104282. [CrossRef]

49. Liu, T.; Zhao, Q.; Xie, Y.; Jiang, D.; Chu, Z.; Jin, W. In situ fabrication of aloe-like Au–ZnO micro/nanoarrays for ultrasensitive biosensing of catechol. *Biosens. Bioelectron.* **2020**, *156*, 112145. [CrossRef]

50. Palomar, Q.; Gondran, C.; Lellouche, J.-P.; Cosnier, S.; Holzinger, M. Functionalized tungsten disulfide nanotubes for dopamine and catechol detection in a tyrosinase-based amperometric biosensor design. *J. Mater. Chem. B* **2020**, *8*, 3566–3573. [CrossRef]

51. Diculescu, V.; Chiorcea-Paquim, A.; Brett, C.M. Applications of a DNA-electrochemical biosensor. *TrAC Trends Anal. Chem.* **2016**, *79*, 23–36. [CrossRef]

52. Wang, X.; Chen, R.; Sun, L.; Yu, Z. Study on the Antioxidant Capacities of Four Antioxidants Based on Oxidizing Guanine in a Composite Membrane. *Int. J. Electrochem. Sci.* **2014**, *9*, 6834–6842.

53. Mello, L.D.; Kubota, L.T. Antioxidant capacity of Ilex paraguariensis extracts by using HRP-based biosensor. *Lat. Am. Appl. Res.* **2014**, *44*, 325–329.

54. Mello, L.D.; Sotomayor, M.D.P.T.; Kubota, L.T. HRP-based amperometric biosensor for the polyphenols determination in vegetables extract. *Sens. Actuators B Chem.* **2003**, *96*, 636–645. [CrossRef]

55. Garcia, L.F.; Benjamin, S.R.; Marreto, R.N.; Lopes, F.M. Laccase Carbon Paste Based Biosensors for Antioxidant Capacity. The Effect of Different Modifiers. *Int. J. Electrochem. Sci.* **2015**, *10*, 5650–5660.

56. Akyüz, E.; Başkan, K.S.; Tütem, E.; Apak, R. Novel Protein-Based Solid-Biosensor for Determining Pro-oxidant Activity of Phenolic Compounds. *J. Agric. Food Chem.* **2017**, *65*, 5821–5830. [CrossRef]

57. Akyüz, E.; Başkan, K.S.; Tütem, E.; Apak, R. Novel Iron(III)−Induced Prooxidant Activity Measurement Using a Solid Protein Sensor in Comparison with a Copper(II)−Induced Assay. *Anal. Lett.* **2020**, *53*, 1489–1503. [CrossRef]

58. Medina-Plaza, C.; De Saja, J.; Rodriguez-Mendez, M.L. Bioelectronic tongue based on lipidic nanostructured layers containing phenol oxidases and lutetium bisphthalocyanine for the analysis of grapes. *Biosens. Bioelectron.* **2014**, *57*, 276–283. [CrossRef]

59. Cetó, X.; Capdevila, J.; Mínguez, S.; Del Valle, M. Voltammetric BioElectronic Tongue for the analysis of phenolic compounds in rosé cava wines. *Food Res. Int.* **2014**, *55*, 455–461. [CrossRef]

60. Cui, C.; Wang, Q.; Liu, Q.; Deng, X.; Liu, T.; Li, D.; Zhang, X. Porphyrin-based porous organic framework: An efficient and stable peroxidase-mimicking nanozyme for detection of H2O2 and evaluation of antioxidant. *Sens. Actuators B Chem.* **2018**, *277*, 86–94. [CrossRef]

61. Becker, M.M.; Ribeiro, E.B.; Marques, P.R.B.D.O.; Marty, J.-L.; Nunes, G.S.; Catanante, G. Development of a highly sensitive xanthine oxidase-based biosensor for the determination of antioxidant capacity in Amazonian fruit samples. *Talanta* **2019**, *204*, 626–632. [CrossRef]

 *biosensors*

Article

Yeast-Based Fluorescent Sensors for the Simultaneous Detection of Estrogenic and Androgenic Compounds, Coupled with High-Performance Thin Layer Chromatography

Liat Moscovici [1], Carolin Riegraf [2,3], Nidaa Abu-Rmailah [1], Hadas Atias [1], Dror Shakibai [1], Sebastian Buchinger [2], Georg Reifferscheid [2] and Shimshon Belkin [1,*]

[1] Department of Plant and Environmental Sciences, Institute of Life Sciences, Hebrew University of Jerusalem, Jerusalem 91904, Israel; liatmosc@savion.huji.ac.il (L.M.); nidaan8@gmail.com (N.A.-R.); hadas.atias@mail.huji.ac.il (H.A.); drors1987@gmail.com (D.S.)

[2] Department Biochemistry, Ecotoxicology, Federal Institute of Hydrology (BfG), Am Mainzer Tor 1, 56068 Koblenz, Germany; carolin.riegraf@gmail.com (C.R.); Buchinger@bafg.de (S.B.); reifferscheid@bafg.de (G.R.)

[3] RWTH Aachen University, Department of Ecosystem Analysis, Worringerweg 1, D-52074 Aachen, Germany

[*] Correspondence: shimshon.belkin@mail.huji.ac.il; Tel.: +972-2-6584192

Received: 15 October 2020; Accepted: 4 November 2020; Published: 8 November 2020

Abstract: The persistence of endocrine disrupting compounds (EDCs) throughout wastewater treatment processes poses a significant health threat to humans and to the environment. The analysis of EDCs in wastewater remains a challenge for several reasons, including (a) the multitude of bioactive but partially unknown compounds, (b) the complexity of the wastewater matrix, and (c) the required analytical sensitivity. By coupling biological assays with high-performance thin-layer chromatography (HPTLC), different samples can be screened simultaneously, highlighting their active components; these may then be identified by chemical analysis. To allow the multiparallel detection of diverse endocrine disruption activities, we have constructed *Saccharomyces cerevisiae*-based bioreporter strains, responding to compounds with either estrogenic or androgenic activity, by the expression of green (EGFP), red (mRuby), or blue (mTagBFP2) fluorescent proteins. We demonstrate the analytical potential inherent in combining chromatographic compound separation with a direct fluorescent signal detection of EDC activities. The applicability of the system is further demonstrated by separating influent samples of wastewater treatment plants, and simultaneously quantifying estrogenic and androgenic activities of their components. The combination of a chemical separation technique with an optical yeast-based bioassay presents a potentially valuable addition to our arsenal of environmental pollution monitoring tools.

Keywords: Bioassays; high performance thin layer chromatography; endocrine disrupting compounds; fluorescent proteins; wastewater

---

1. Introduction

Endocrine disrupting chemicals (EDCs) are exogenous agents with structural similarity to endogenous hormones, that may therefore interfere with natural hormonal activity by blocking, competing or mimicking natural hormones [1]. The biological effects associated with exposure to EDCs include numerous physiological processes, among them homeostasis disruption, immunological damages and developmental impairments [2]. Furthermore, some EDCs were suggested to act as carcinogenic agents [3]. These adverse health effects, in some cases triggered by exposure to parts-per-billion level concentrations [4–6], have raised concern among public health authorities. This concern is exacerbated when considering the possible manifestation of certain health effects associated with exposure to EDCs across generations [2]. Besides being a human health risk, the release of EDCs to aquatic systems also poses a significant ecological threat. Continuous exposure to such chemicals may affect local species and ecosystems. It has been shown, for example, that chronic exposure to EDCs causes abnormalities in the reproductive system of certain fish species [1,7–9].

The estrogen (ER) and the androgen (AR) receptors are prominent members of a hormone receptor superfamily that mediates a wide range of significant biological activities. These vary from reproductive development to the regulation of the cardiovascular system, the immune system, the central nervous system, and more [10,11]. Both of these steroid hormone receptors are comprised of three main structural domains with similar functionalities: (a) the N-terminal transcription regulation domain; (b) the ligand binding domain (LBD), which attaches to the target ligand, prompting a conformational change that allows the receptor–ligand complex to enter the nucleus; and (c) the DNA binding domain. A ligand–receptor dimer complex is translocated into the nucleus, and binds to a DNA consensus sequence at the promoter of a target gene, known as the Hormone Response Element (HRE), leading to the transcription of the gene [2,10,12].

Detecting specific pollutants that exhibit hormonal activity in complex environmental samples, e.g., treated wastewater, is very challenging due to the complexity of the matrix. Such samples may contain a large variety of EDCs, as well as numerous unknown EDC metabolites, which may also exert endocrine disrupting activity. The need to focus on the presence of unknown but bioactive compounds, renders traditional detection methodologies, e.g., liquid- or gas-chromatography coupled to mass spectrometry (LC/MS and GC/MS, respectively), less suitable for the detection of EDCs in complex samples [12,13].

A possible alternative to such methods is the use of whole-cell biosensors, genetically engineered to emit a detectable signal upon exposure to chemicals exerting hormonal activity. Such effect-based methods require no prior information regarding the chemical structure of the EDCs in the sample. When applied in a microtiter plate-based assay, this methodology allows quantifying the combined biological effect resulting from exposure to the EDCs in the sample, rather than assessing the concentrations of individual chemicals. This feature adds important information on the toxic effects of compound mixtures. A notable disadvantage of this approach is the inability to distinguish between different components exerting similar biological effects [14]. To overcome this difficulty, a number of reports describing the direct coupling of effect-based methods to high-performance thin-layer chromatography (HPTLC) have been published recently [13,15,16]

The combination of chemical separation by high-performance thin-layer chromatography (HPTLC) and the effect-based assay by yeast-based sensor strains allowed the separation of environmental samples and the discovery of individual sample components exhibiting hormonal activity. These active compounds can then be removed from the HPTLC plate and identified via traditional analytical methods (e.g., LC/MS, GC/MS) [17,18]. However, since a separate assay has to be conducted for each biological endpoint, the throughput potential of this approach is low. In an answer to this need, the present article describes the development of yeast (*Saccharomyces cerevisiae*)-based sensors that detect the presence of chemicals exerting androgenic and estrogenic activity by expressing spectrally different fluorescent proteins. Following characterization of the constructed sensor strains in a 96-well microtiter plate format, they were sprayed over HPTLC plates, in which model compounds and later wastewater samples were separated. Following incubation, EDCs with estrogenic and androgenic activities were simultaneously detected in the same sample. In contrast to a previously described *Arxula adeninivorans*-based assay with similar objectives [19], we have employed a spray-on-technology to apply a uniform layer of the yeast bioreporters to the HPTLC surface [20]. This methodology allows the control of the thickness of the suspension layer and produces clear and sharp bands, as opposed to an immersion procedure.

2. Materials and Methods

2.1. Chemicals

Testosterone (CAS: 58-22-0) and 5α-androstan-17β-ol-3-one (DHT, CAS: 521-18-6) were used as androgenic reference compounds. Estrone (E1, CAS: 53-16-7), 17β-estradiol (E2, CAS: 50-28-2), estriol (E3, CAS: 50-27-1), and 17α-ethinylestradiol (EE2, CAS: 57-63-6), were used as estrogenic reference compounds. These chemicals were of the highest analytical grade and were purchased from Sigma-Aldrich. Stock solutions of reference compounds (0.5 mg/mL for DHT, 5 mg/mL for the rest) were prepared in ethanol. Chromatographic separation was performed on silica gel HPTLC plates of type 60G F_{254} (20 × 10 cm or 10x10 cm) purchased from Merck. Solvents used for HPTLC were of the highest analytical grade and were purchased from Merck.

2.2. Yeast Strains, Plasmids, and Growth Conditions

Two previously constructed *S. cerevisiae* sensor strains and two plasmids were employed in this study as a basis for the construction of the new fluorescent bioreporters. The two strains, harboring either the human estrogen nuclear receptor (hER) or the human androgen receptor (hAR), integrated into the yeast genome, were purchased from BioTech (Knoxville, TN, USA). The two plasmids were kindly donated by Prof. S. Ripp (University of Tennessee, Knoxville, TN, USA). Plasmids pUTK407 and pUTK420 [21,22], contained an estrogenic or an androgenic hormone response element (HRE), respectively, between bidirectional constitutive and strong yeast promoters, upstream of the *luxA* and *luxB* genes of the luminescent bacterium *Photorhabdus luminescens*. Plasmid pUTK407 carries two copies of the human estrogen HRE, located between the constitutive divergent promoters GPD and ADH1. Similarly designed, pUTK420 carries four copies of the human androgen HRE, located between the same two promoters. The *luxA* and *luxB* genes are located downstream of the GPD and ADH1 promoters, respectively. The palindromic nature of the HRE region forms a hairpin structure that represses the activation of the GPD and ADH1 promoters. Upon binding of the ligand–receptor complex to its respective HRE, this hairpin structure is released, and both *luxA* and *luxB* are divergently transcribed, yielding the two structural subunits of the bacterial luciferase.

In the present study, the *luxB* sequences from the skeleton plasmids [21,22] were replaced by one of three fluorescent protein (FP) genes (Figure 1): green (EGFP), red (mRuby2), or blue (mTagBFP2). These genes were extracted from plasmids pFA6a-link-yoEGFP-spHIS5, pFA6a-link-yomRuby2-spHIS5 and pFA6a-link-yomTag BFP2-spHIS5, respectively [23]. All three plasmids were a kind gift from Wendell Lim and Kurt Thorn (Addgene plasmids #44,838, #44,858 and #44,836, respectively). A set of complementary oligonucleotides, for each of the fluorescent protein sequences, was obtained using KOD Hot Start DNA polymerase (Merck). The six new plasmids generated in this manner are listed in Table 1, and the PCR primers employed for cloning the fluorescent protein genes are listed in Table S1. DNA manipulations were performed according to standard protocols [24].

Yeast extract–peptone–dextrose (YPD) liquid medium was used for routine growth of the plasmid-free *S. cerevisiae* strains. A modified minimal medium without uracil (Sigma-Aldrich/ Formedium, United Kingdom) was employed to grow strains harboring plasmids with a uracil selective marker.

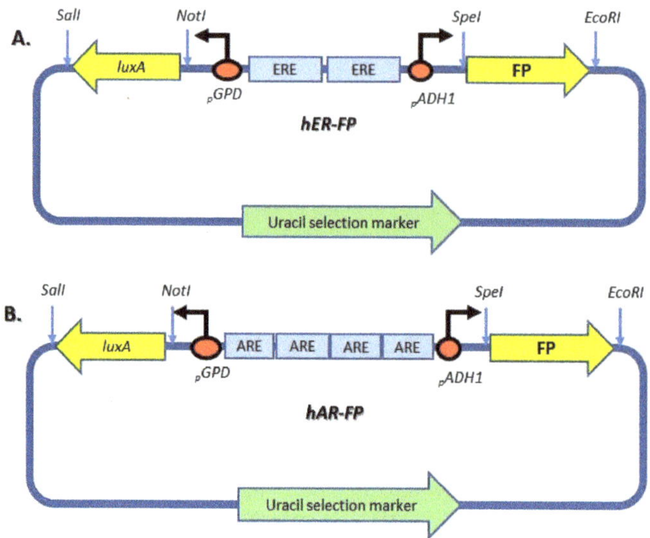

Figure 1. Schematic design of the hER-FP (**A**) and hAR-FP (**B**) plasmids used in this study for the detection of estrogenic and androgenic activity, respectively. Plasmid hER-FP (**A**), derived from plasmid pUTK407 (21), contains two copies of the human estrogen response element (ERE). Plasmid hAR-FP (**B**), derived from plasmid pUTK420 (22), contains four copies of the human androgen response element (ARE). Upon binding of a receptor–ligand complex to its respective response element, a hairpin structure is released and activation of the constitutive GPD and ADH1 promoter is enabled, resulting in transcription of *luxA* and fluorescent genes (FP), either EGFP, BFP, or Ruby.

Table 1. *Saccharomyces cerevisiae* strains and plasmids used in this study.

Strain or Plasmid	Description	Source or Reference
	***S. cerevisiae* parental strains**	
hER	MATa; leu2; his3; Human estrogen receptor gene in the chromosome.	[25]
hAR	BJ 1991 MATa; prb1-1122; pep4-3; leu2; trp1; ura3-52; GAL Human estrogen receptor gene in the chromosome.	[26]
	Parental plasmids	
pFA6a-link-yomRuby2/yomTagBFP2/yoEGFP	containing Ruby, BFP, EGFP gene respectively	Addgene #44858 #44839#44836 respectively [23]
pUTK407	Contains the *luxA* and *luxB* genes expressed from the bidirectional promoters GPD and ADH1 separated with two EREs.	[21]
pUTK420	Contains the *luxA* and *luxB* genes expressed from the bidirectional promoters GPD and ADH1 separated with four AREs.	[22]
	ER and AR reporter plasmids	
ER fluorescent reporter (FP)	pUTK407 in which *luxB* was substituted by EGFP/Ruby/BFP gene.	This study
AR fluorescent reporter (FP)	pUTK420 in which *luxB* was substituted by EGFP/Ruby/BFP gene.	This study

Table 1. *Cont.*

Strain or Plasmid	Description	Source or Reference
Fluorescent sensor strains		
hER-EGFP	Contains the EGFP gene contred by ADH1 promoter and the *luxA* gene expressed from GPD promoter, with two repeats of EREs.	This study
hER-Ruby	Contains the Ruby gene contred by ADH1 promoter and the *luxA* gene expressed from GPD promoter, with two repeats of EREs.	This study
hER-BFP	Contains the BFP gene contred by ADH1 promoter and the *luxA* gene expressed from GPD promoter, with two repeats of EREs.	This study
hAR-EGFP	Contains the EGFP gene contred by ADH1 promoter and the *luxA* gene expressed from GPD promoter, with two repeats of EREs.	This study
hAR-Ruby	Contains the Ruby gene contred by ADH1 promoter and the *luxA* gene expressed from GPD promoter, with two repeats of EREs.	This study
hAR-BFP	Contains the BFP gene contred by ADH1 promoter and the *luxA* gene expressed from GPD promoter, with two repeats of EREs.	This study

Transformation of plasmids into *S. cerevisiae* cells was performed according to a standard lithium acetate protocol [27]. Briefly, *S. cerevisiae* cells were grown overnight (30 °C, 200 rpm), and then diluted 100-fold into 10 mL of fresh medium. The cells were grown under the same conditions to late exponential growth phase, until the optical density at 600 nm (OD_{600}) was 0.6 to 1, and then washed with nuclease-free water, and pelleted at 10,600 RCF (Eppendorf 5417C) at room temperature. Following a second washing step with 0.1 M lithium acetate (CAS: 6108-17-4, Sigma-Aldrich), the cells were resuspended in 240 μL 50% polyethylene glycol (PEG 4000, CAS: 25322-68-3, Merck). Subsequently, 36 μL of 1 M lithium acetate, 25 μL of carrier DNA (10 mg/mL, fish testes denatured DNA, CAS: 100403-24-5, USA Bioworld) and 45 μL of the DNA to be transformed (up to 1 μg) were added. The cells were then incubated at 30 °C on a gently rotating platform for 45 min, following which they were subjected to a 25 min heat shock at 42 °C. The cells were pelleted, resuspended in 50 μL of nuclease-free water, and plated on minus *ura* synthetic complete (SC) agar plates [28]. Successful transformations were verified by colony PCR and sequencing.

2.3. Endocrine Assay in 96-Well Plates

Yeast strains were grown overnight (30 °C, 250 rpm) in a selective medium (a synthetic complete medium, lacking uracil, unless mentioned otherwise). The culture was diluted 100-fold in fresh medium, and re-grown under the same conditions to late exponential growth phase (O.D.600 = 0.6–1). Aliquots (40 μL) of the culture were then dispensed into each well of a 96-well black clear-bottom microtiter plate (Greiner), containing 80 μL of reference compounds at predetermined concentrations (0.0122–200 μg/l). The reference compounds, either E2 (estradiol), or testosterone, were dissolved in ethanol, which also served as a negative control (1%).

The 96-well plates were incubated at 30 °C for 18 h ± 1 h in a TECAN plate reader (Infinite M200 PRO), and the fluorescent signal was read every hour, following a 10 sec vigorous shaking of the plate. The readings were performed using excitation/emission wavelengths of 559/600 nm for Ruby, 488/507 nm for EGFP and 399/454 nm for BFP. Fluorescence values are displayed as the instrument's arbitrary relative fluorescence units (RFU).

2.4. Calculation of the Corrected Fluorescence and the Reporter Gene Induction in 96-Well Plates

A corrected fluorescence [13] value, accounting for cell density as well as background fluorescence, was calculated according to the following equation:

$$F_c(i) = \frac{\left[A_{Fluorescence}(i) - \overline{B_{fluorescence}}(i) \right]}{\left[A_{600}(i) - \overline{B_{600}}(i) \right]}$$

where $Fc(i)$ is corrected fluorescence for test i (sample dilutions, reference dilutions, negative control); $A_{fluorescence}(i)$ is fluorescence intensity for test i; $\overline{B_{fluorescence}}(i)$ is mean fluorescence intensity for blank replicates of test i; $A_{600}(i)$ is absorbance at 600 nm for test i; and $\overline{B_{600}}(i)$ is absorbance at 600 nm for blank replicates of test i.

2.5. Endocrine Assay on the HPTLC Plate Surface

HPTLC plates (Silica gel, F_{254}, 20 × 10 cm, Merck) were developed with methanol to 5 mm below the rim, dried at 120 °C for 30 min, and stored in a desiccator at room temperature until used. Samples and reference compounds were applied by an Automatic TLC Sampler 4 (ATS 4, CAMAG, Muttenz, Switzerland), in amounts ranging from 0.5 pg to 1000 pg per spot, as described before [13]. Samples were focused with 100% methanol to a distance of 20 mm, followed by 5 min drying in a chemical hood. Chromatographic development, up to 10 mm below the rim, was performed using an Automated Multiple Development System (AMD 2, CAMAG, Muttenz, Switzerland). For the separation of estrogen-like compounds, a chloroform/acetone/petroleum ether (55:20:25) mixture was used as the mobile phase. Ethylacetate/n-hexane (50:50) served as the mobile phase for the separation of androgen-like compounds [13,29]. For the simultaneous detection of estrogenicity and androgenicity, a mobile phase consisting of ethylacetate/n-hexane (50:50) was employed. Following separation, the plates were dried in a chemical hood until the organic solvents evaporated [13].

For the detection of endocrine activity on the HPTLC plate, an overnight culture of the yeast-based bioreporters was centrifuged at 10,600 RCF for 5 min (Eppendorf centrifuge 5417C). The pellet was then resuspended in fresh minimal medium without uracil, and regrown under the same conditions to late exponential growth phase (OD_{600} = 0.6–1). The cells were sprayed homogenously on the developed HPTLC plate, either manually with a glass reagent sprayer (CAMAG, Muttenz, Switzerland), or by using an automated spraying device (CAMAG Derivatizer, CAMAG, Muttenz, Switzerland, 2.5 mL, spraying level 3, yellow nozzle). Images of the fluorescent signal were obtained after an incubation of 4 h to 18 h at 30 °C in an opaque plastic box, in which humidity was maintained by a water-soaked paper towel. The fluorescent EGFP signals were detected using Fusion FX imaging system (Vilber Lourmat) at excitation and emission wavelength of 365 nm 565 nm, respectively. The fluorescent Ruby and BFP signals were detected using a TLC Scanner 4 (CAMAG) operated under the *visionCATS* software (version 2.3. SP1, CAMAG, Muttenz, Switzerland). Ruby signals were detected at λ_{ex} = 525 nm with a cutoff filter of 540 nm, and BFP signals at λ_{ex} = 396 nm and a cutoff filter of 400 nm. Additionally, qualitative assessment was performed on images acquired with a TLC Visualizer 2 (CAMAG) operated under the *visionCATS* software (version 2.3. SP1,) under long wavelength UV light (λ_{em} = 366 nm).

2.6. Preparation of Wastewater Samples

Freshly collected influent samples of municipal wastewater treatment plants were centrifuged (Thermo Scientific, Sorvall RC 6 Plus Centrifuge, 17,000 RCF, 20 min) and the supernatant was filtered through a glass fiber filter (Pall, type A/C, Ø 47 µm). Filtered samples were concentrated by solid-phase extraction (SPE) using Oasis HLB cartridges (200 mg, 6 mL). The columns were conditioned by the successive application of 2 mL n-heptane, 2 mL acetone, three aliquots of 2 mL methanol, and four aliquots of 2 mL deionized water. Methanol (2 × 4 mL) was used to elute the adsorbed sample components from the cartridges. The extracts were reduced to 500 µL using a Turbo Vap II Concentration Workstation (Biotage AB, Uppsala, Sweden) under a gentle nitrogen flow. The extracts were transferred into amber glass vials. The evaporation tubes were rinsed three times with methanol, which was then used to fill up the extracts to a final volume of 1 mL, resulting in a final 200-fold enrichment. The extracts were stored at −20 °C until use.

2.7. Fluorescent Microscopy

Microscope images were obtained using a VF1200 confocal microscope (Olympus, Tokyo, Japan) with a 60 × 1.42 oil objective. The excitation wavelengths and emission filters were 488/507 nm for EGFP, 559/600 nm for Ruby and 399/454 nm for BFP.

2.8. Data Processing and Statistical Analysis

Statistical analysis of the HPTLC combined bioassays was performed as previously described [13], using the intensity (in arbitrary units) of the determined peak areas obtained with the *visionCATS* software (version 2.5. SP1, CAMAG). Data were further processed using Excel® and R version 3.5.2 (R Core Team, Vienna, Austria) [30], the 'drc' [31] and the 'ggplot' [32] packages. Signal-to-noise-ratios (S/N) were determined to calculate the limit of detection (LOD) and limit of quantification (LOQ) with S/N ≥ 3 and S/N ≥ 10, respectively.

3. Results

3.1. Sensor Strain Characterization in a 96-Well Plate Assay

To reach the study's objective, developing a method for multi parallel detection of different endocrine disruptors in environmental samples within a single assay, we have designed and constructed a battery of yeast-based sensor strains. The six members of the sensor panel, addressing two target chemical groups (with estrogenic or androgenic activities), with three reporter proteins (green, red and blue) each, are listed in Table 1. The responses to their designated model targets have first been characterized in a conventional 96-well microtiter plate procedure in liquid medium. The activities of two bioreporters out of this list, an estrogenic sensor (ER-Ruby) and an androgenic (AR-BFP) are presented, as an example, in Figure 2.

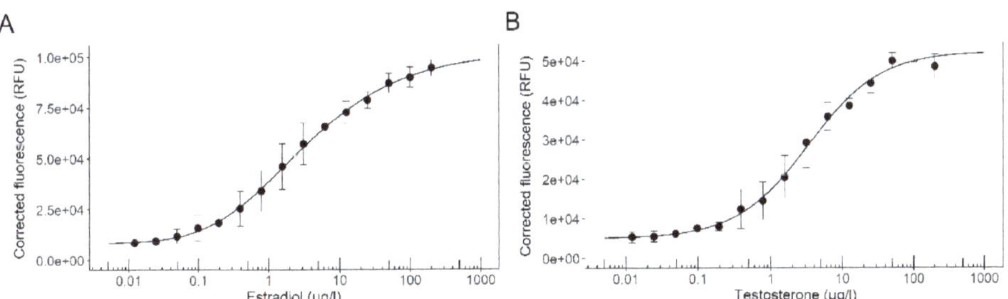

Figure 2. Dose-response curves of two fluorescent bioreporters. (**A**) Response of strain ER-Ruby to β-Estradiol (E2; 0.0122–200 µg/L); (**B**) Response of strain AR-BFP to testosterone (0.0122–200 µg/L). Estrogenic and androgenic activities were determined after an incubation time of 18 h at 30 °C. Corrected fluorescence values were calculated as detailed in Materials and Methods (Section 2.4), and the error bars show the respective standard errors. The solid line was fitted to the data using a five-parameter log-logistic function.

The response of both strains to increasing concentrations of their respective reference compounds followed a classic asymmetrical logistic dose-response curve. Both strains exhibited high sensitivity towards their reference compounds; calculated LOD values for 17β-estradiol (E2) by ER-Ruby and testosterone by AR-BFP were 0.032 and 0.070 µg/L, respectively.

Thin-layer chromatography allows the separation of compounds based on polarity-governed partitioning between the solid and the mobile phases. A planar yeast estrogen screen (pYES) to investigate estrogenic activities of environmental sample components was previously described, using yeast bioreporters coupled with HPTLC [29,33,34]. Recently, Riegraf et al. [13] demonstrated that additional modes of action can be addressed by combining other yeast-based reporter gene assays with HPTLC, including androgenic effects. In all of these cases, β-galactosidase has been employed as the reporter entity, and only one type of endocrine activity could be assayed in a single plate. In the present study, the optimized working protocols described in these publications have been used to test the performance of the newly developed fluorescent yeast strains. Mixtures of estrogenic [E1 (estrone), E2 (17β- estradiol), EE2 (17α- ethinylestradiol) and E3 (estriol)] and androgenic [DHT (dihydrotestosterone) and testosterone] reference compounds were applied and subsequently separated by HPTLC, as described under Material and Methods. The yeast fluorescent bioreporter cells were then sprayed as a thin layer on top of the silica plate, and their activity was monitored following an 18 h incubation. Successful performance was achieved for both the ER-Ruby and the AR-BFP strains. The resulting fluorescence signals of the ER-Ruby strain are shown in Figure 3, and the corresponding dose-response curve in Figure 4. The respective results for the AR-BFP strain are displayed in Figures 5 and 6.

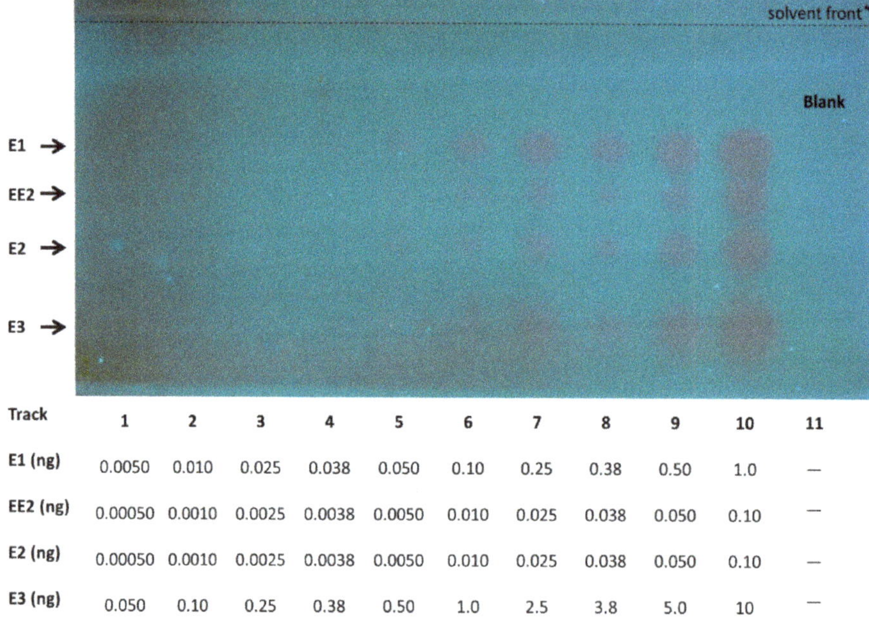

Track	1	2	3	4	5	6	7	8	9	10	11
E1 (ng)	0.0050	0.010	0.025	0.038	0.050	0.10	0.25	0.38	0.50	1.0	—
EE2 (ng)	0.00050	0.0010	0.0025	0.0038	0.0050	0.010	0.025	0.038	0.050	0.10	—
E2 (ng)	0.00050	0.0010	0.0025	0.0038	0.0050	0.010	0.025	0.038	0.050	0.10	—
E3 (ng)	0.050	0.10	0.25	0.38	0.50	1.0	2.5	3.8	5.0	10	—

Figure 3. Detection of estrogenic activity by the ER-Ruby sensor strain following high-performance thin-layer chromatography (HPTLC) separation. Different amounts (indicated below the image) of a mixture consisting of the reference compounds estrone (E1), 17α-ethinylestradiol (EE2), 17β-estradiol (E2), and estriol (E3) were separated by a two-step chromatographic development using methanol and a chloroform/ethyl acetate/petroleum ether mixture (55:20:25, *v/v/v*). Ethanol served as blank on track 11. Following an 18 h incubation at 30 °C, fluorescence was imaged using a TLC Visualizer 2 at λ_{ex} = 366 nm. Signals were enhanced using the enhancement tool of the *visionCATS* software.

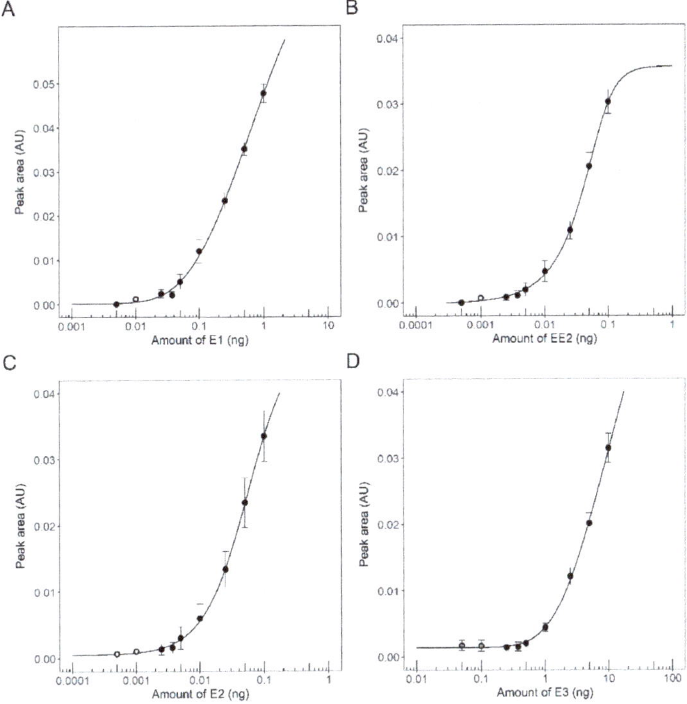

Figure 4. Dose-response curves of estrogenic reference compounds Estrone (E1, **A**), 17α-Ethinylestradiol (EE2, **B**), 17β-Estradiol (E2, **C**), and Estriol (E3, **D**), as derived from the experiment described in Figure 3 above. Fluorescent signal intensity determined by the peak area was plotted against the applied amount. Data points represent the mean values of the signal and the error bars show the respective standard error (n_{black} = 3, n_{grey} = 2 and n_{white} = 1). The solid lines were fitted to the data using a five-parameter log-logistic function.

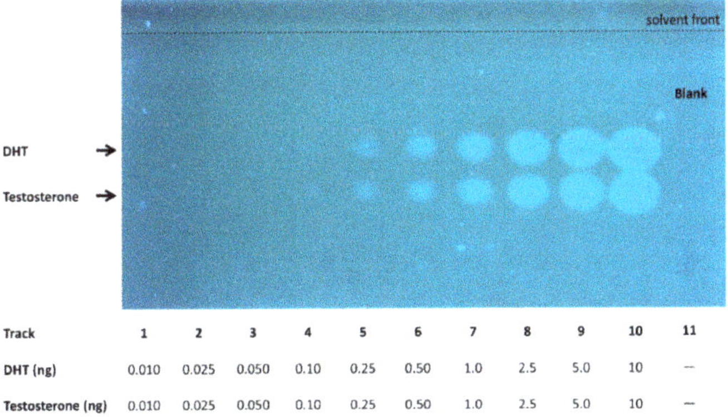

Track	1	2	3	4	5	6	7	8	9	10	11
DHT (ng)	0.010	0.025	0.050	0.10	0.25	0.50	1.0	2.5	5.0	10	—
Testosterone (ng)	0.010	0.025	0.050	0.10	0.25	0.50	1.0	2.5	5.0	10	—

Figure 5. Detection of androgenic activity by the AR-BFP sensor strain following HPTLC separation. Different amounts (indicated below the image) of a testosterone/DHT (5α-androstan-17β-ol-3-one) mixture were separated by a two-step chromatographic development using methanol and an ethyl

acetate/n-hexane mixture (50:50, *v/v*). Ethanol served as blank on track 11. Following an 18 h incubation at 30 °C, fluorescence was imaged using a TLC Visualizer 2 at λ_{ex} = 366 nm. Signals were enhanced using the enhancement tool of the *visionCATS* software.

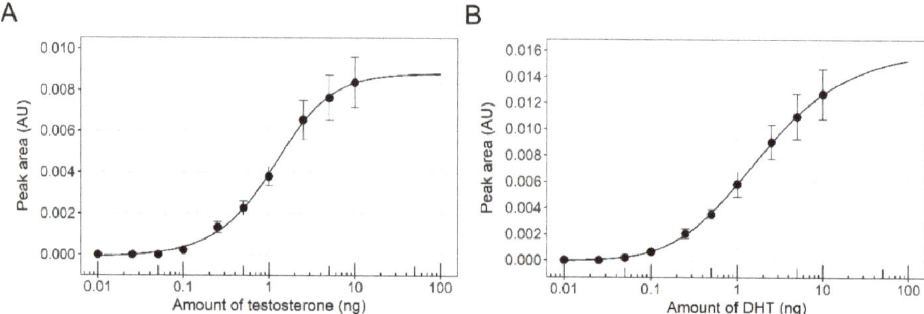

Figure 6. Dose-response curves of androgenic reference compounds testosterone (**A**) and 5α-Androstan-17β-ol-3-one (DHT, **B**), as derived from the experiment described in Figure 5 above. Fluorescent signal intensity determined by the peak area detected using a TLC Scanner 4 (λ_{ex} = 396 nm, cut-off filter of 400 nm) was plotted against the applied amount. Data points represent the mean values of the signal, and the error bars show the respective standard error (n = 3). The solid lines were fitted to the data using a five-parameter log-logistic function.

3.2. Simultaneous Detection of Estrogenic and Androgenic Activities—Model Compounds

To demonstrate the simultaneous detection of different EDC classes, we have tested the response of a 1:1 blend (v:v) of two of the newly constructed yeast fluorescent bioreporters, ER-Ruby (red fluorescence) and AR-EGFP (green fluorescence), to a E2/testosterone mixture. Exposure was first conducted in liquid culture; Figure 7 presents microscopic images obtained in the presence of different concentrations of the model inducers. Both strains have responded by a bright fluorescence of the respective newly synthesized reporter protein, with a larger number of fluorescent cells visible in the presence of the higher inducer dose.

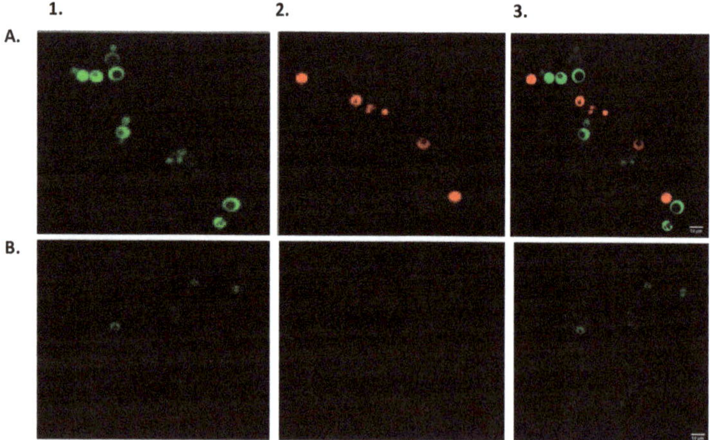

Figure 7. Simultaneous detection of estrogenic (E2) and androgenic (testosterone) model compounds by mixed ER-Ruby and AR-EGFP sensor strains in liquid culture. Images were obtained with a VF1200 confocal microscope (Olympus, Tokyo, Japan) with a 60 × 1.42 oil objective, following an 18 h incubation

at 30 °C. Images were taken in a sequential mode using λ_{ex} = 488 nm and λ_{em} = 505–540 nm for the EGFP signal (column 1) and λ_{ex} = 561 nm and λ_{em} = 570λ620 nm for Ruby (column 2). Merged images are shown in column 3. Row (**A**)—a testosterone/E2 mixture, 250 ng/l each; Row (**B**)—no ligand.

Following the successful demonstration of the combined fluorescent detection of the two hormonal activities in liquid medium, we have investigated the possibility of combining the two assays also on the surface of an HPTLC plate. To characterize a possible crosstalk between the quantification of estrogenic and androgenic effects, two reporter strains (ER-Ruby and AR-BFP) were sprayed, either individually or in a 1:1 mixture, onto an HPTLC plate on which two blends of estrogenic and androgenic model compounds were chromatographically separated, either individually or in a mixture. The on-plate responses of the two reporter strains are presented in Figure 8. On all three plates, an estrogen mix consisting of E1, E2 and E3 and an androgenic mix consisting of testosterone and DHT were applied on tracks 1–3 and 4–6, respectively, while a combined mixture of both classes was applied on tracks 7–9. Activity detection was performed either with strain AR-BFP alone (top panel), ER-Ruby alone (middle panel) and with a mixture of the two reporter strains (bottom panel). The HPTLC plates were scanned at the wavelengths appropriate for the excitation of both fluorescent reporter proteins, and the fluorescence intensities are represented by the heights of the bars in Figure 8. These data were also used to calculate the LOD values for the different compounds, as presented in Table 2.

Table 2. Limit of detection (LOD), limit of quantification (LOQ) and Rf values calculated for the ER–Ruby and AR-BFP sensor strains in response to HPTLC-separated reference compounds.

		Single Strain		Both Strains		Rf Values
		Individual	Mix	Individual	Mix	
E3	Mean LOD (ng)	0.82	1.43	1.7	3.6	0.19
	SE (ng)	0.02	0.09	0.4	1.9	
E2	Mean LOD (ng)	0.0081	0.012	0.03	0.04	0.60
	SE (ng)	0.0009	0.001	0.02	0.02	
E1	Mean LOD (ng)	0.0120	0.016	0.05	0.06	0.76
	SE (ng)	0.0002	0.002	0.03	0.04	
	Mean LOQ (ng)	0.042	0.058			
	SE (ng)	0.003	0.008			
Testosterone	Mean LOD (ng)	0.3	0.70	0.8 (*)	0.9	0.46
	SE (ng)	0.2	0.04	0.2	0.2	
DHT	Mean LOD (ng)	<0.5	<0.5	0.3 (*)	0.4	0.61
	SE (ng)			0.1	0.1	

(*) Two replicates only.

The blue fluorescence sensor strain AR-BFP detected the two androgenic model chemicals (Figure 8A), when on their own (tracks 4–6) as well as in the presence of the estrogenic mixture (tracks 7–9). Similarly, the red fluorescence sensor ER-Ruby displayed a clear response (Figure 8B) to the three estrogenic compounds, both in the absence (tracks 1–3) and the presence (tracks 7–9) of the androgenic substances. In all cases, the responses were dose-dependent. The intensity of the responses when both hormone classes were combined (Figure 8C) tended to be lower than when the mixtures were separated. This was more evident in the case of the ER-Ruby sensor (Figure 8B), which also displayed a minor response to the higher doses of the androgenic compounds. The reduced sensitivities in the combined presence of the two hormone classes are also evident from the higher LODs listed in Table 2.

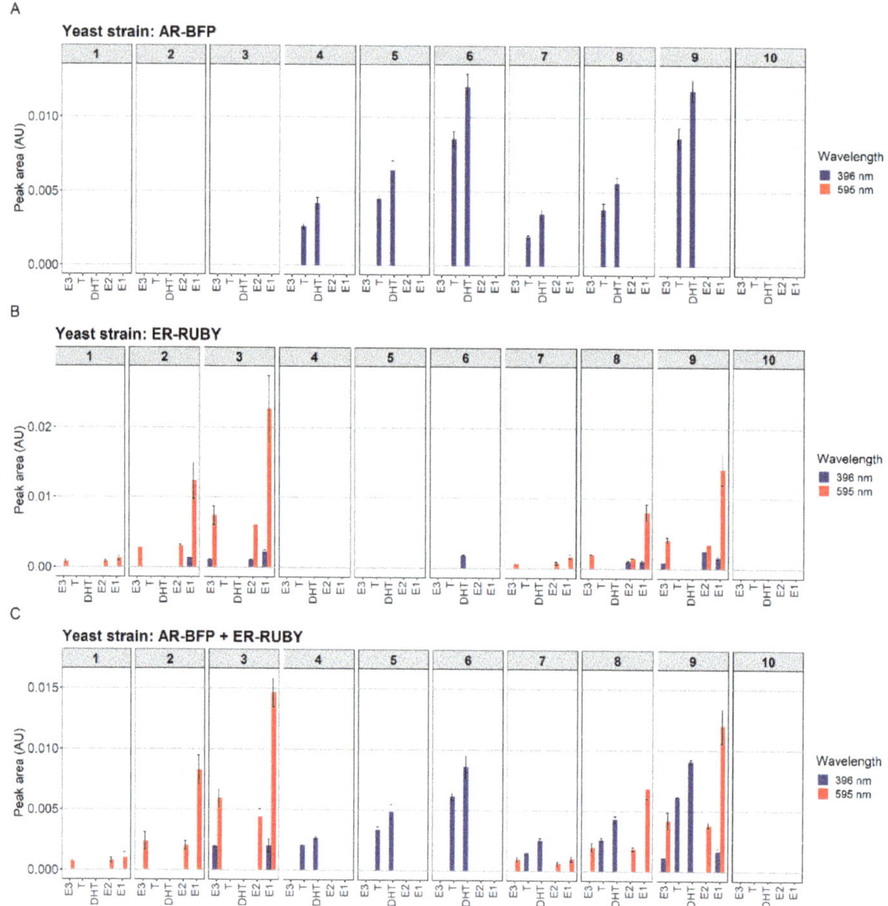

Figure 8. Multi-effect detection of chromatographically separated model compound mixtures by strain AR-BFP (**A**), strain ER-Ruby, (**B**) and both strains together (**C**). The HPTLC plates with the sprayed-on sensor cells were incubated for 18 h at 30 °C, following which fluorescence was scanned at λ = 396 nm and λ = 525 nm. Fluorescence intensities are presented in arbitrary units (AU). The estrogen mixture, separated on tracks 1–3 and 7–9, was composed of E1 (0.01 ng, 0.05 ng and 0.1 ng), E2 (0.005 ng, 0.01 ng and 0.02 ng), and E3 (0.5 ng, 1 ng and 2 ng). The androgen mixture, separated on tracks 4–6 and 7–9, was composed of testosterone (T; 0.5 ng, 1 ng and 5 ng) and DHT (0.5 ng, 1 ng and 5 ng). Ethanol served as blank on track 10. A two-step chromatographic development was performed, with 100% methanol for the first 20 mm, and an ethyl acetate/n-hexane 1:1 mixture for the next 70 mm.

3.3. Simultaneous Detection of Estrogenic and Androgenic Activities—Wastewater Samples

The robustness of the HPTLC combined bioassay using the newly developed fluorescent yeast sensor cells was further demonstrated by its application to wastewater samples. Concentrated influent samples from two municipal wastewater treatment plants (WWTPs) were separated by HPTLC, following which the plates were sprayed with a 1:1 mixture of two reporter strains, ER-Ruby and AR-BFP, and incubated for 18 h at 30 °C. An estrogenic mix consisting of E1, E2, and E3 and an androgenic mix consisting of DHT and testosterone were identically treated. Figure 9 displays an image of the plate (top) and the fluorescence scans of the individual lanes (bottom).

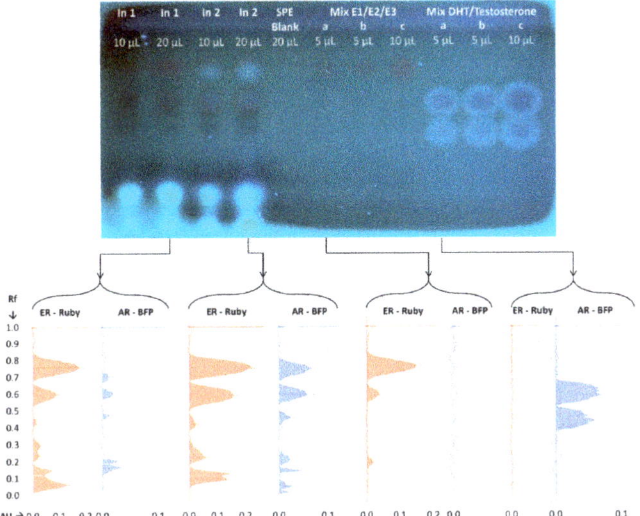

Figure 9. Simultaneous detection of androgenic and estrogenic effects in extracts of two wastewater treatment plants (WWTP) influent (In1, In2; 10 µl and 20 µl each) by a 1:1 mix of two yeast bioreporter strains, ER-Ruby and AR-BFP. An estrogen mix consisting of E1 (a: 0.1, b: 0.2 and c: 0.4 ng), E2 (0, 20 and 40 pg) and E3 (1, 2 and 4 ng), and an androgen mix of DHT (a: 10, b: 25 and c: 50 ng) and testosterone (10, 25 and 50 ng) were separated on the same plate. Ethanol served as blank on track 10. A two-step chromatographic development was performed, with 100% methanol for the first 20 mm, and an ethyl acetate/n-hexane 1:1 mixture for the next 70 mm. Top: plate image displaying the fluorescent signal using a TLC Visualizer 2 at λ_{ex} = 366 nm following an 18 h exposure at 30 °C. Signals were enhanced using the enhancement tool of the *visionCATS* software. Bottom: scans of the individual tracks for both ER-Ruby- and AR-BFP-fluorescence (TLC Scanner settings: Ruby: λ_{ex} = 525 nm, a 540 nm cutoff filter; BFP: λ_{ex} = 396 nm a 400 nm cutoff filter).

As is evident from the data in Figure 9, both influent samples contain components with potential estrogenic and androgenic activities. For example, the Ruby fluorescent signal, detected in both influent samples at Rf = 0.75, is similar in its chromatographic migration distance to that of E1, hinting at the presence of a molecule with an estrogenic activity chemically resembling E1. In the same location in sample In2 there is also an apparent androgenic activity, which is not shared by sample In1. Additional fluorescent signals were detected at Rf = 0.6 in sample In 2, and to a lesser extent also in In1 for both yeast bioreporter strains. These signals showed a similar migration behavior as the estrogenic model compound E2 and the androgenic model compound DHT. Clearly, however, while the fluorescent intensity of each active sample constituent on the HPTLC plate can be accurately quantified, a determination of its actual concentration cannot be performed before it is fully identified by analytical chemical means.

4. Discussion

Cell-based assays utilizing reporter gene technology have been widely promoted for environmental monitoring; in contrast to chemical analysis, they provide information about the biological effects of a tested sample, even if its exact composition is unknown. Furthermore, such assays can be tailored to detect chemicals with specific modes of action, such as an interference with cellular processes, including hormone receptor signaling [35,36].

Many assays for the detection of endocrine disrupting effects employ yeast cells as the cellular chassis, and rely on the expression of *LacZ* as the reporting element [25,26,37,38]. However, to measure the level of *LacZ* expression, the cell membrane must be disrupted, and an external substrate needs to be added to visualize the extent of gene expression induced by the target chemical(s). In contrast, fluorescent proteins provide a fast and accurate alternative [39,40] for signal detection even in living cells. Furthermore, as demonstrated in the present article, it allows the multi-parallel detection of several types of fluorescent proteins, differing in their optical attributes. Here we have described a set of *Saccharomyces cerevisiae* fluorescent protein-based bioreporters for the detection of two classes of endocrine disrupting chemicals in complex samples, in combination with sample separation by HPTLC, with LOD values similar or lower to those reported for other cellular assays (Table 2).

Probably the most relevant comparison of our results is to those described in the recently published reports of an *Arxula Adeninivorans*-based cell assay, employing three fluorescent bioreporter strains for the detection of estrogenic, androgenic, and progesterone activity, combined with HPTLC [19]. In that study, a clear peak with the fluorescent *Arxula adeninivorans* yeast strain G1212/YRC102-hER-DsRed, was detected for E2 > 0.0075 ng, without chromatographic separation. This result indicates a similar sensitivity compared to the strain generated in the current study, hER-Ruby, with a LOD of 0.008 ng, after chromatographic separation. The minimal detectable amount of DHT by the YRC102-hAR-GFP strain was 0.025 ng [19], lower than our strain hAR-BFP with a LOD < 0.5 (following chromatographic separation). However, a direct comparison of the performance characteristics of the bioreporter strains described by Chamas et al. [19] is not possible, since in the current study the cells were applied to the plate surface by spraying, compared to immersion of the developed plate in a suspension of the yeast bioreporters. The spray-on technology, as described herein, allows control over the thickness of the yeast layer as well as the plate's moisture, producing clear and sharp bands [20]. Furthermore, as noted above, the LOD values reported here were determined following full chromatographic migration and separation.

A main objective of the current study was to allow a multi-parallel effect detection of both estrogenic and androgenic activities by complex samples and their components. This was achieved by combining sample separation by HPTLC with the concomitant application of two representative sensor strains; in these, red and blue fluorescence are induced in the presence of estrogenic or androgenic compounds, respectively. The general functionality of both reporters in parallel is demonstrated in Figures 8 and 9. The former figure describes a systematic experiment addressing the possible crosstalk between the assays, in terms of false positive signals and effects on assay sensitivity. In the case of the AR-BFP strain, intensity of the signal caused by the androgenic model compounds remained virtually unchanged in the presence of the estrogenic model compounds. Furthermore, no false positive results were detected at 595 nm, indicating that the potential presence of estrogenic compounds is detected specifically by the ER-Ruby strain even in the presence of androgenic compounds. A possible crosstalk between androgenic compounds such as DHT and an activation of the estrogen receptor is discussed in literature either via a direct binding of DHT to the estrogen receptor [36] or due to a metabolic conversion of testosterone [41,42], but was not found in the present study.

Nevertheless, apparent false positive signals were detected at 396 nm as a strain ER-Ruby response, on its own, to estrogenic compounds. The increasing intensity of this signal correlates with the increasing signals at 595 nm in the presence of higher concentrations of the model compounds. Possibly, when the concentration of this fluorescent protein is sufficiently high, the low excitation of Ruby at 396 nm becomes visible. This artefact may be avoided by the introduction of a better-tuned filter system. The 595 nm signal intensity, indicating the presence of the Ruby protein, is lower when the estrogenic and androgenic model compounds are applied in a mixture. While actual signal intensities remain quite stable in the co-exposure (Figure 8), assay sensitivity is reduced due to an increased noise level leading to reduced S/N-values. While future refinement of the optical system is certainly desirable, the results presented in Figure 8 clearly demonstrate the possibility of a specific multi-parallel effect detection.

The specificity and robustness of the constructed yeast strains and the adaptability of the combined method for screening environmental samples is demonstrated by the characterization of influent samples from two different WWTPs. As in any chromatographic system, TLC separation of a complex sample into its components depends on adsorption properties of the compounds to the solid phase, as well as their solubility and migration distance with the mobile phase. Since estrogens and androgens share similar structures and physico-chemical properties, it is challenging to separate the two classes from each other with HPTLC [19]. The use of the multi-parallel effect detection is thus advantageous, as different effects may be detected even if the active compounds are not fully separated. This is clearly visible for the influent sample from the second WWTP (sample In 2). In this sample, two strong signals for estrogenicity and androgenicity at Rf-values of 0.75 and 0.61 are superimposed. The estrogenic bands at Rf 0.75 and 0.61 migrate similarly to E1 and E2, respectively; the androgenic compound Rf migrates similarly to DHT, while no candidate compound can be assigned to the androgenic signal at Rf = 0.75. Interestingly, no androgenic signals above a signal to noise ratio of 3 could be detected in the influent sample of WWTP 1 at these positions, underlining the specificity of the detection even in complex environmental samples.

5. Conclusions

We have presented a panel of sensitive, specific, and robust fluorescent bioreporter *S. cerevisiae* strains for the detection of compounds exerting estrogenic and androgenic effects in complex samples, and have demonstrated their efficacy in the analysis of both model compounds and multi-component wastewater samples. The use of fluorescent proteins as reporter elements obviates the requirement for cell lysis, substrate addition, and a second incubation step. The information emerging from such assays, when combined with thin layer chromatographic separation, can serve to restrict subsequent chemical analysis only to the small number of active fractions of the sample; this; in turn, will allow significant savings in time and costs, and a more focused and efficient risk assessment. To further broaden the applicability of the approach, it would be desirable to expand the panel of sensor strains for additional hormone classes, lower limits of detection, and optimize separation conditions for different classes of endocrine and endocrine-like compounds. Furthermore, as for other whole-cell based assays, the approach described above is limited to the detection of compounds that are at least partially permeable into the intracellular environment; enhancing cellular permeability to a broader spectrum of potential target molecules should therefore be another objective in the design of future EDC sensor strains.

Supplementary Materials: The following are available online at http://www.mdpi.com/2079-6374/10/11/169/s1, Table S1: Oligonucleotide primer sequences used for the construction of the plasmids in this study.

Author Contributions: Conceptualization, L.M., S.B. (Sebastian Buchinger), G.R. and S.B.; Methodology, L.M., C.R., and S.B. (Sebastian Buchinger); Investigation, L.M., C.R., N.A.-R., H.A. and D.S.; Resources, G.R., S.B. (Sebastian Buchinger) and S.B.; Data Curation, C.R.; Writing—Original Draft Preparation, L.M., C.R; Writing—Review & Editing, S.B. (Sebastian Buchinger) and S.B.; Visualization, L.M. and C.R.; Supervision, G.R., S.B. (Sebastian Buchinger) and S.B.; Project Administration, L.M.; Funding Acquisition, G.R., S.B. (Sebastian Buchinger) and S.B. All authors have read and agreed to the published version of the manuscript.

Funding: This work was supported by the Federal Ministry of Education and Research (BMBF), Karlsruhe Project Management Agency, Germany (Grant No. 02WIL1387), and the Ministry of Science, Technology and Space (MOST), Israel (Grant No. WT1402/2560) in the framework of the German–Israeli Cooperation in Water Technology Research. Work at the Hebrew University was also supported in part by the Minerva Center for Bio-Hybrid Complex Systems.

Acknowledgments: The authors thank Marina Ohlig and Ramona Pfänder for excellent technical assistance. The help of Naomi Melamed-Book (Bioimaging unit, Alexander Silberman Institute of life Sciences, The Hebrew University of Jerusalem) with the confocal microscopy is gratefully acknowledged.

Conflicts of Interest: The authors declare no conflict of interest.

References

1. Da Silva, A.P.A.; De Oliveira, C.D.L.; Quirino, A.M.S.; da Silva, F.D.M.; de Aquino Saraiva, R.D.; Silva-Cavalcanti, J.S. Endocrine disruptors in aquatic environment: Effects and consequences on the biodiversity of fish and amphibian species. *Aquat. Sci. Technol.* **2018**, *6*, 35–51. [CrossRef]
2. Kerdivel, G.; Habauzit, D.; Pakdel, F. Assessment and molecular actions of endocrine-disrupting chemicals that interfere with estrogen receptor pathways. *Int. J. Endocrinol.* **2013**, *2013*, 1–14. [CrossRef] [PubMed]
3. Soto, A.M.; Sonnenschein, C. Environmental causes of cancer: Endocrine disruptors as carcinogens. *Nat. Rev. Endocrinol.* **2010**, *6*, 363–370. [CrossRef] [PubMed]
4. Vandenberg, L.N.; Colborn, T.; Hayes, T.B.; Heindel, J.J.; Jacobs, D.R., Jr.; Lee, D.H.; Shioda, T.; Soto, A.M.; vom Saal, F.S.; Welshons, W.V.; et al. Hormones and endocrine-disrupting chemicals: Low-dose effects and nonmonotonic dose responses. *Endocr. Rev.* **2012**, *33*, 378–455. [CrossRef] [PubMed]
5. Colborn, T.; Saal, F.S.V.; Soto, A.M. Developmental effects of endocrine-disrupting chemicals in wildlife and humans. *Environ. Health Perspect.* **1993**, *101*, 378–384. [CrossRef]
6. Casals-Casas, C.; Desvergne, B. Endocrine disruptors: From endocrine to metabolic disruption. *Annu. Rev. Physiol.* **2011**, *73*, 135–162. [CrossRef] [PubMed]
7. Arlos, M.J.; Parker, W.J.; Bicudo, J.R.; Law, P.; Hicks, K.A.; Fuzzen, M.L.; Andrews, S.A.; Servos, M.R. Modeling the exposure of wild fish to endocrine active chemicals: Potential linkages of total estrogenicity to field-observed intersex. *Water Res.* **2018**, *139*, 187–197. [CrossRef]
8. Svenson, A.; Allard, A.S.; Ek, M. Removal of estrogenicity in Swedish municipal sewage treatment plants. *Water Res.* **2003**, *37*, 4433–4443. [CrossRef]
9. Sumpter, J.P.; Johnson, A.C. 10th Anniversary Perspective: Reflections on endocrine disruption in the aquatic environment: From known knowns to unknown unknowns (and many things in between). *J. Environ. Monit.* **2008**, *10*, 1476–1485. [CrossRef]
10. Davey, R.A.; Grossmann, M. Androgen receptor structure, function and biology: From bench to bedside. *Clin. Biochem. Rev.* **2016**, *37*, 3–15. [PubMed]
11. Yaşar, P.; Ayaz, G.; User, S.D.; Güpür, G.; Muyan, M. Molecular mechanism of estrogen-estrogen receptor signaling. *Reprod. Med. Biol.* **2017**, *16*, 4–20. [CrossRef] [PubMed]
12. Campana, C.; Pezzi, V.; Rainey, W.E. Cell-based assays for screening androgen receptor ligands. *Semin. Reprod. Med.* **2015**, *33*, 225–234. [CrossRef] [PubMed]
13. Riegraf, C.; Reifferscheid, G.; Belkin, S.; Moscovici, L.; Shakibai, D.; Hollert, H.; Buchinger, S. Combination of yeast-based in vitro screens with high-performance thin-layer chromatography as a novel tool for the detection of hormonal and dioxin-like compounds. *Anal. Chim. Acta* **2019**, *1081*, 218–230. [CrossRef]
14. Adeniran, A.; Sherer, M.; Tyo, K.E.J. Yeast-based biosensors: Design and applications. *FEMS Yeast Res.* **2015**, *15*, 1–15. [CrossRef]
15. Bergmann, A.J.; Simon, E.; Schifferli, A.; Schönborn, A.; Vermeirssen, E.L.M. Estrogenic activity of food contact materials—Evaluation of 20 chemicals using a yeast estrogen screen on HPTLC or 96-well plates. *Anal. Bioanal. Chem.* **2020**, *412*, 4527–4536. [CrossRef]
16. Shakibai, D.; Riegraf, C.; Moscovici, L.; Reifferscheid, G.; Buchinger, S.; Belkin, S. Coupling high-performance thin-layer chromatography with bacterial genotoxicity bioreporters. *Environ. Sci. Technol.* **2019**, *53*, 6410–6419. [CrossRef] [PubMed]
17. Klingelhöfer, I.; Morlock, G.E. Sharp-bounded zones link to the effect in planar chromatography-bioassay-mass spectrometry. *J. Chromatogr. A* **2014**, *1360*, 288–295. [CrossRef]
18. Klingelhöfer, I.; Morlock, G.E. Bioprofiling of surface/wastewater and bioquantitation of discovered endocrine-active compounds by streamlined direct bioautography. *Anal. Chem.* **2015**, *87*, 11098–11104. [CrossRef] [PubMed]
19. Chamas, A.; Pham, H.T.M.; Jähne, M.; Hettwer, K.; Gehrmann, L.; Tuerk, J.; Uhlig, S.; Simon, K.; Baronian, K.; Kunze, G. Separation and identification of hormone-active compounds using a combination of chromatographic separation and yeast-based reporter assay. *Sci. Total. Environ.* **2017**, *605*, 507–513. [CrossRef]

20. Schoenborn, A.; Schmid, P.; Bräm, S.; Reifferscheid, G.; Ohlig, M.; Buchinger, S. Unprecedented sensitivity of the planar yeast estrogen screen by using a spray-on technology. *J. Chromatogr. A* **2017**, *1530*, 185–191. [CrossRef] [PubMed]

21. Sanseverino, J.; Gupta, R.K.; Layton, A.C.; Patterson, S.S.; Ripp, S.A.; Saidak, L.; Simpson, M.L.; Schultz, T.W.; Sayler, G.S. Use of Saccharomyces cerevisiae BLYES Expressing Bacterial Bioluminescence for Rapid, Sensitive Detection of Estrogenic Compounds. *Appl. Environ. Microbiol.* **2005**, *71*, 4455–4460. [CrossRef]

22. Eldridge, M.L.; Sanseverino, J.; Layton, A.C.; Easter, J.P.; Schultz, T.W.; Sayler, G.S. Saccharomyces cerevisiae BLYAS, a new bioluminescent bioreporter for detection of androgenic compounds. *Appl. Environ. Microbiol.* **2007**, *73*, 6012–6018. [CrossRef]

23. Lee, S.; Lim, W.A.; Thorn, K.S. Improved blue, green, and red fluorescent protein tagging vectors for S. cerevisiae. *PLoS ONE* **2013**, *8*, e67902. [CrossRef]

24. Green, M.R.; Sambrook, J. *Molecular Cloning: A Laboratory Manual*; Cold Spring Harbor Laboratory Press: New York, NY, USA, 2012; Volume 1.

25. Routledge, E.J.; Sumpter, J.P. Estrogenic activity of surfactants and some of their degradation products assessed using a recombinant yeast screen. *Environ. Toxicol. Chem.* **1996**, *15*, 241–248. [CrossRef]

26. Purvis, I.J.; Chotai, D.; Dykes, C.W.; Lubahn, D.B.; French, F.S.; Wilson, E.M.; Hobden, A.N. An androgen-inducible expression system for Saccharomyces cerevisiae. *Gene* **1991**, *106*, 35–42. [CrossRef]

27. Gietz, R.D.; Woods, R.A. Transformation of yeast by lithium acetate/single-stranded carrier DNA/polyethylene glycol method. *Methods Enzymol.* **2002**, *350*, 87–96. [CrossRef]

28. Dymond, J.S. Saccharomyces cerevisiae growth media. *Methods Enzymol.* **2013**, *533*, 191–204. [CrossRef]

29. Spira, D.; Reifferscheid, G.; Buchinger, S. Combination of high-performance thin-layer chromatography with a specific bioassay—A tool for effect-directed analysis. *J. Planar Chromatogr. Mod. TLC* **2013**, *26*, 395–401. [CrossRef]

30. R Core Team. *R: A Language and Environment for Statistical Computing*; R Foundation for Statistical Computing: Vienna, Austria, 2013. Available online: http://www.R-project.org/ (accessed on 11 May 2020).

31. Ritz, C.; Baty, F.; Streibig, J.C.; Gerhard, D. Dose-response analysis using R. *PLoS ONE* **2015**, *10*, e0146021. [CrossRef]

32. Wickham, H. *Ggplot2: Elegant Graphics for Data Analysis*; Springer: New York, NY, USA, 2016. Available online: https://ggplot2.tidyverse.org/authors.html (accessed on 11 May 2020).

33. Buchinger, S.; Spira, D.; Bröder, K.; Schlüsener, M.; Ternes, T.; Reifferscheid, G. Direct coupling of thin-layer chromatography with a bioassay for the detection of estrogenic compounds: Applications for effect-directed analysis. *Anal. Chem.* **2013**, *85*, 7248–7256. [CrossRef]

34. Dausend, C.; Weins, C.; Frimmel, F. A New bioautographic screening method for the detection of estrogenic compounds. *Chromatographia* **2004**, *60*, 207–211. [CrossRef]

35. Jarosova, B.; Érseková, A.; Hilscherová, K.; Loos, R.; Gawlik, B.M.; Giesy, J.P.; Bláha, L. Europe-wide survey of estrogenicity in wastewater treatment plant effluents: The need for the effect-based monitoring. *Environ. Sci. Pollut. Res.* **2014**, *21*, 10970–10982. [CrossRef]

36. Leusch, F.D.; Chapman, H.F.; Heuvel, M.R.V.D.; Tan, B.L.; Gooneratne, S.R.; Tremblay, L.A. Bioassay-derived androgenic and estrogenic activity in municipal sewage in Australia and New Zealand. *Ecotoxicol. Environ. Saf.* **2006**, *65*, 403–411. [CrossRef]

37. Coldham, N.G.; Dave, M.; Sivapathasundaram, S.; McDonnell, D.P.; Connor, C.; Sauer, M.J. Evaluation of a recombinant yeast cell estrogen screening assay. *Environ. Health Perspect.* **1997**, *105*, 734–742. [CrossRef] [PubMed]

38. Miller III, C.A.; Tan, X.; Wilson, M.; Bhattacharyya, S.; Ludwig, S. Single plasmids expressing human steroid hormone receptors and a reporter gene for use in yeast signaling assays. *Plasmid* **2010**, *63*, 73–78. [CrossRef]

39. Bovee, T.F.; Helsdingen, R.J.R.; Hamers, A.R.M.; Van Duursen, M.B.M.; Nielen, M.W.F.; Hoogenboom, R.L.A.P. A new highly specific and robust yeast androgen bioassay for the detection of agonists and antagonists. *Anal. Bioanal. Chem.* **2007**, *389*, 1549–1558. [CrossRef]

40. Bovee, T.F.; Helsdingen, R.J.; Koks, P.D.; Kuiper, H.A.; Hoogenboom, R.L.; Keijer, J. Development of a rapid yeast estrogen bioassay, based on the expression of green fluorescent protein. *Gene* **2004**, *325*, 187–200. [CrossRef]

41. Handa, R.J.; Pak, T.R.; Kudwa, A.E.; Lund, T.D.; Hinds, L. An alternate pathway for androgen regulation of brain function: Activation of estrogen receptor beta by the metabolite of dihydrotestosterone, 5α-androstane-3β,17β-diol. *Horm. Behav.* **2008**, *53*, 741–752. [CrossRef]
42. Beresford, N.; Routledge, E.J.; Harris, C.A.; Sumpter, J.P. Issues arising when interpreting results from an in vitro assay for estrogenic activity. *Toxicol. Appl. Pharmacol.* **2000**, *162*, 22–33. [CrossRef] [PubMed]

Publisher's Note: MDPI stays neutral with regard to jurisdictional claims in published maps and institutional affiliations.

Article

Electromagnetic Piezoelectric Acoustic Sensor Detection of Extracellular Vesicles through Interaction with Detached Vesicle Proteins

Loránd Románszki [1], Zoltán Varga [1], Judith Mihály [1], Zsófia Keresztes [1,*] and Michael Thompson [2]

[1] Institute of Materials and Environmental Chemistry, Research Centre for Natural Sciences, H-1117 Budapest, Hungary; romanszki.lorand@ttk.hu (L.R.); varga.zoltan@ttk.hu (Z.V.); mihaly.judith@ttk.hu (J.M.)

[2] Department of Chemistry, University of Toronto, Toronto, ON M5S 3H6, Canada; m.thompson@utoronto.ca

* Correspondence: keresztes.zsofia@ttk.hu

Received: 2 October 2020; Accepted: 9 November 2020; Published: 11 November 2020

Abstract: An electromagnetic piezoelectric acoustic sensor (EMPAS) was used to study the non-specific adsorption of human red blood cell-derived extracellular vesicle preparations. Vesicle storage history (temperature and duration) highly affected the obtained results: The signal change, namely the frequency decrease of the crystal measured at 20 °C, was negligibly small (<1 s^{-2}) when the vesicle solutions had previously been stored at 4 °C, and was in the order of 10 s^{-2} when the vesicle solutions had been stored at −30 °C. Moreover, the rate of frequency decrease increased exponentially with the storage time at −30 °C. Upon a 4 °C storage period following the −30 °C storage period of the same sample, the measured frequency decrease dropped, suggesting a partial relaxation of the system. The results are explained by the disintegration of the vesicles triggered by the freeze–thaw cycle, likely due to the detachment of proteins from the vesicle surface as was proved by size-exclusion chromatography. Surface modification of the sensor crystal provided the possibility of signal enhancement, as the maximum rate of the frequency change for the same vesicle concentrations was higher on hydrophobic, octadecyl trichlorosilane–modified quartz than on hydrophilic, bare quartz. The EMPAS signal has been associated with the amount of detached proteins, which in turn is proportional to the originating vesicle concentration.

Keywords: electromagnetic piezoelectric acoustic sensor; quartz; adsorption; diagnostics; extracellular vesicle

1. Introduction

Extracellular vesicles (EVs) are cell-released lipid particles containing a large variety of proteins, nucleic acids and metabolites. They offer vast biological information as a cargo in intercellular communication, and are also a potential target in diagnostic techniques [1,2]. In order to use EVs in diagnosis, well characterized sampling and separation requirements must be fulfilled.

Guideline literature MISEV2018 [3] stresses that purity requirements must be met in order for EV isolates to provide reliable information on associated functional activity. Isolation of EVs from non-EV material and soluble non–EV-associated proteins is of high importance as the impurities may interfere with particle number counts and biomarker analysis.

Problems concerning EV particle detection methods originate from the fact that most of the methods cannot distinguish EVs from other biological nanoparticles and may require antibody labelling or antibody capture. One possible solution is to consider indirect indicators, such as total protein-to-particle or protein-to-lipid ratios. Particle number determination methods include flow cytometry [4,5], nanoparticle tracking analysis [6,7], and resistive pulse sensing [8–10].

Surface-based analytical methods, such as optical surface plasmon resonance [11–13] can provide sensitive particle counting. Acoustic surface sensitive detection techniques do not require an optically transparent medium; therefore, the analysis of biological fluids is more feasible. Changes in conditions at the interface between the sensor and the liquid due to molecular/cellular adsorption/desorption or viscosity change all shift the frequency of the oscillating sensor that provides the measurement signal. The oscillating sensor surface can be modified to specifically or non-specifically capture EVs. A quartz crystal microbalance (QCM) with a CD63 EV specific antibody modified sensor has been used to detect EVs from isolates [14].

The electromagnetic piezoelectric acoustic sensor (EMPAS) [15,16] operates by remotely induced ultra-high frequency (up to 1.06 GHz) acoustic shear wave generation in the AT-cut quartz sensor crystal. The EMPAS is a highly sensitive method due to the applicable ultra-high frequency overtones. As the EMPAS, unlike the QCM, does not require electrode connections, the homogenous surface of the quartz sensor allows for versatile surface chemistry routes to engineer the surface for a given interaction. EMPAS devices have been used for detection of HIV-2 antibodies in human serum [17], endotoxin of pathogenic *E. coli* in full human blood plasma [18], cocaine [19], breast and prostate cancer metastasis biomarker (PTHrP) [20]. They have also been used in the study of antifouling coating against undiluted goat serum [21], full human blood plasma [22], human serum and bovine milk [23], as well as in the determination of plasmin enzyme of pM concentration levels [24].

The main aim of this study was to investigate the ability of an EMPAS device to detect EVs. For this purpose, a well characterized [25,26] model EV was used: namely, a red blood cell-derived EV. We aimed to follow the non-specific adsorption; therefore, no specific immuno-capture capability has been introduced. Only a clear comparison of interactions with hydrophobic and hydrophilic surfaces has been probed. The results have revealed that a hydrophobic surface has higher binding capacity than a hydrophilic surface in the assayed EV dispersion. A huge increase in the adsorption capacity was observed after one freeze–thaw cycle of the EV sample. This observation shed light not only on the effect of freezing on the integrity of EVs, but also to the possible application of EMPAS devices for the characterization of the purity of EV samples without the need for separation.

2. Materials and Methods

2.1. Isolation of Red Blood Cell Derived EVs (REVs)

The use of human blood samples was carried out by following the guidelines and regulations of the Helsinki Declaration in 1975, and was approved by the Scientific and Research Ethics Committee of the Hungarian Medical Research Council (ETT TUKEB 6449-2/2015). Human red blood cell-derived EVs (REVs) were isolated according to the protocol described previously [25]. Briefly, 3×6 mL whole blood was collected from healthy volunteers with informed consent into vacuum tubes containing EDTA anticoagulant (VACUETTE® TUBE 6 mL K3EDTA, Greiner Bio_One, Mosonmagyaróvár, Hungary). The red blood cells (RBC) were sedimented by centrifugation at 2500 *g* for 15 min (Nüve NF800R), and after the removal of plasma and white blood cells, were resuspended in isotonic saline solution (0.9% NaCl, B. Braun AG) and washed three times (2500 *g* for 10 min at 4 °C). 3.5 mL of washed RBC were diluted to 10 mL with phosphate buffered saline (PBS, pH = 7.4, Sigma-Aldrich and stored at 4 °C for 7 days. REVs were isolated from the RBC suspension via two consecutive centrifugations (1500 *g* for 10 min and 2850 *g* for 30 min). The supernatant was centrifuged at 16,000 *g* for 30 min (Eppendorf 5415R). The final pellet was suspended in total 100 µL PBS, and purified with size exclusion chromatography (SEC) using a 3.5 mL gravity column filled with Sepharose CL-2B gel (GE Healthcare, Sweden). The 100 mL EV sample was pipetted into the column followed by 900 mL PBS. The second 1 mL fraction containing the purified EVs was collected and kept at 4 °C.

2.2. Microfluidic Resistive Pulse Sensing (MRPS)

MRPS is based on the Coulter principle realized in a microfluidic cartridge. MRPS measurements were performed with an nCS1 instrument (Spectradyne LLC, Torrance, CA, USA). The samples were diluted 20-fold with a bovine serum albumin (BSA, Sigma-Aldrich, Hungary) solution at 1 mg/mL in PBS buffer (Sigma-Aldrich, Hungary), filtered through an Amicon Ultra 0.5 centrifugal filter with 100 kDa MWCO (Merck-Millipore, Hungary) according to the manufacturer's instructions. All measurements were performed using factory calibrated TS-400 cartridges with a measurement range from 65 nm to 400 nm.

2.3. Fourier-Transform Infrared Spectroscopy (FTIR)

Attenuated total reflection (ATR) infrared spectroscopy was used to obtain signature spectra for protein and lipid composition of isolated REV samples. 3 µL of fresh REV sample was spotted and dried under ambient conditions on the top of the diamond ATR element of a single reflection Golden Gate accessory (Specac Ltd., Orpington, UK) fitted into a Varian 2000 FT-IR spectrometer (Varian Inc., Paolo Alto, CA, USA). Spectra were recorded by coaddition of 64 individual scans with a nominal spectral resolution of 2 cm^{-1}. Before the spectrum evaluation, ATR correction and the subtraction of the PBS buffer background were performed. For all spectral manipulations, the GRAMS/32 software package (Galactic Inc, Birmingham, AL, USA) were used.

2.4. Size Exclusion Chromatography with On-Line Fluorescence Detection (Flu-SEC)

Flu-SEC was used to study the release of soluble proteins from EVs upon a freeze–thaw cycle. 10 µL of REV sample was injected into a Jasco HPLC system (Jasco, Tokyo, Japan) consisting of a PU-2089 pump with a UV-2075 UV/Vis detector and a FP-2020 fluorescence detector controlled by the Chromnav software v. 1.17.02. Tricorn 5/100 glass columns (GE Healthcare Bio-Sciences AB) were packed with Sepharose CL-2B (GE Healthcare Bio-Sciences AB), and the eluent was PBS with a flow rate of 0.5 mL/min. The fluorescence chromatograms were collected at excitation and emission wavelength corresponding to the intrinsic fluorescence of proteins (280/340 nm), and the area under the curve of the EV peak and the free protein peak was used to quantify the amount of unbound proteins.

2.5. Electromagnetic Piezoelectric Acoustic Sensor (EMPAS) Measurements

Hydrophilic (θ_a = 15°) [24] quartz crystal discs were obtained using the following method. The discs underwent ultrasound cleaning in detergent solution for 30 min in test tubes, then were copiously rinsed with tap water and distilled water, followed by a 30 min treatment in Piranha solution (3:1 *V/V* mixture of 98% H_2SO_4 and 30% H_2O_2) at 90 °C pre-heated in a water bath. They were then thoroughly rinsed with distilled water and methanol, followed by sonication in another portion of methanol for 2 min, and a final rinse with methanol. Hydrophobic (θ_a = 106°) [24] discs were obtained by further processing the hydrophilic ones as follows. The cleaned, hydrophilic discs were individually transferred into glass vials, which were subsequently placed in an oven maintained at 150 °C for drying. After 2 h, the vials were immediately transferred into a humidity chamber (70–80% relative humidity, room temperature) for 36 h of surface moisturization. Then, the quartz discs were individually transferred to silanized test tubes and moved into a glove box under nitrogen atmosphere. Portions of 1 mL of octadecyltrichlorosilane (OTS) solution in anhydrous toluene (1 µL OTS/999 µL toluene, corresponding to 2.5 mM) were added individually to the test tubes. The tubes were then sealed with rubber stoppers, removed from the glove box, and allowed to stay on a shaker for 2 h. The quartz discs were then rinsed thoroughly with toluene, followed by chloroform, and dried under a gentle stream of nitrogen before being stored in scintillation vials.

The experiments were run using a home-built electromagnetic piezoelectric acoustic sensor [15]. The setup consisted of a Plexiglas flow-through cell (~78 µL internal volume); a ~5 mm diameter hand-wound coil of a 105 µm diameter polyurethane-coated copper wire (Goodfellow); a frequency generator (Hewlett Packard 8648B); a trimmer (muRata, Seminole, FL, USA) placed in parallel with the coil terminals in order for the coil to be tuned to electrical resonance; a lock-in amplifier (SR510, Stanford Research Systems); a diode detector (made in house) to measure the voltage drop developed across the coil terminals at acoustic resonance, with the output being fed into the lock-in amplifier; a digital oscilloscope (Tektronix TDS 210); and a syringe pump (Harvard Apparatus Pump 11) equipped with a 60 mL plastic syringe (Henke-Sass, Wolf GmbH, Germany) providing a flow rate of 25 μL/min. Teflon tubing (1.58 mm od, 0.8 mm id, Supelco, Bellefonte, PA, USA) was used to connect the syringe, flow-through cell and sample vial. In order to minimize temperature effect on the resonant frequency, AT-cut quartz crystal wafer resonators were used. The crystals ($\varnothing$ = 13 mm diameter, t = 83 μm thickness, Laptech Precision Inc., Bowmanville, ON, Canada) were operated at the 49th overtone of their f_0 = 20 MHz fundamental frequency ($f_{49} \approx$ 984 MHz). A code running under LabView 6.0 was used to control the frequency generator and the lock-in amplifier, as well as for data acquisition. REV samples were used in half logarithmic serial dilutions with PBS (pH = 7.4, Sigma-Aldrich), with the dilution factor denoted with D, where $D \equiv -\log (c/c_0)$ = [0.5, 1.0, 1.5, 2.0, 2.5, 3.0], c being the concentration of the diluted solution and c_0 the original concentration of the sample. Measurements were conducted at 20 °C with REV samples stored at 4 °C after 1 h waiting for thermal equilibration or with REV samples stored at 4 °C, followed by a storage at −30 °C for periods ranging from 1 h to several days, after 1 h waiting for thermal equilibration.

3. Results and Discussion

3.1. Characterization of EV Samples

Characterization of REVs has been accomplished by MRPS and FTIR methods. The characteristic results are shown in Figure 1. Figure 1A illustrates MPRS measurement results showing that the REVs, prepared as described, are a homogeneous EV sample with well-defined particle size distribution. The results can be well fitted with a Gaussian distribution which results in a mean diameter of (196.9 ± 0.3) nm with standard deviation (32.26 ± 0.3) nm (adj. R^2 = 0.985). The measured concentration is (1.44 ± 0.01) × 10^{10} mL^{-1} over a size range from 65 nm to 400 nm. The FTIR spectrum in Figure 1B indicates the characteristic lipid and protein bands of REVs. Characteristic bands of the peptide backbones of proteins report at 3299 cm^{-1} (amide A, N–H stretching vibrations), 1654 cm^{-1} (amide I, primarily the C=O stretching vibrations of the amide groups) and 1545 cm^{-1} (amide II, combination of N–H bending and C–N stretching of the amide groups). The long acyl chains of lipids exhibit C–H stretching vibrations in the 3020–2800 cm^{-1} wavenumber region with peaks at 2925 and 2849 cm^{-1}, corresponding to antisymmetric and symmetric stretching of methylene groups, respectively. The relative weak band at 1739 cm^{-1} belongs to the glycerol carbonyl stretching of the phospholipids. Since the protein-to-lipid ratio can be used to assess EV type and purity [3,27], we calculated the spectroscopic protein-to-lipid ratio [28]. Based on the integrated area of amide I band and C–H stretching bands (3020–2800 cm^{-1}) a spectroscopic protein-to-lipid ratio value of 1.52 ± 0.05 was obtained, which is in line with our previous results on purified REV samples [25]. The calculated protein-to-lipid ratio indicates a homogeneous EV sample without impurities. Moreover, as the intensity of the amide I band is proportional with the number of peptide groups, an estimation of the total protein concentration of intact EVs, based on IR spectrum, is also feasible. Using the protocol elaborated by Szentirmai et al. [26], the IR spectroscopy-based total protein concentration of the REV sample is (0.95 ± 0.09) mg/mL.

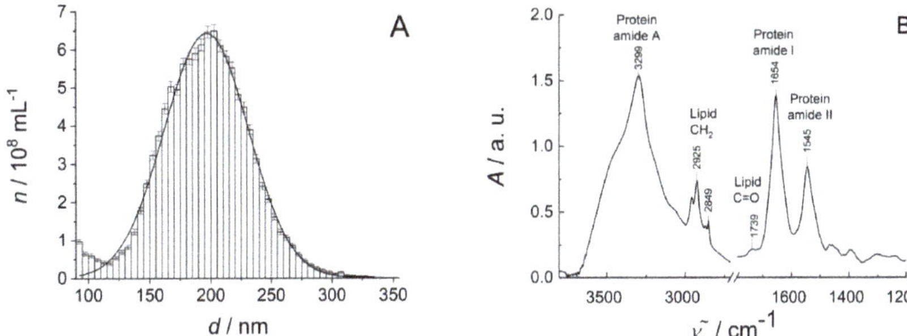

Figure 1. Characterization of red blood cell-derived EVs. (**A**) Particle size distribution of REVs by Microfluidic Resistive Pulse Sensing (MRPS). Number concentration (n) of vesicles as a function of the diameter (d) (50-fold dilution of the original sample); (**B**) Fourier-transform infrared (FTIR) spectrum of the REV sample and the assignment of the main lipid and protein bands.

3.2. Effect of Freezing Temperature and the Duration of Frozen Storage of EVs on EMPAS Signal

Two typical EMPAS measurements are presented in Figure 2. In both cases, after recording the baseline with PBS only, the EV solution enters the cell at $\Delta t = 0$ s, producing a frequency shift of the quartz resonator. The general observation was that the time course of the frequency shift was of a modest rate for EV samples stored at room temperature and 4 °C, even at the relatively high EV concentrations (3.16× dilution of the original sample, $D = 0.5$), whereas the frequency decreased much faster if an identical EV sample had previously been stored at −30 °C, suggesting that freezing might induce structural changes of the EV samples resulting in either an increased number, or a reduced size (i.e., larger diffusion coefficient) of any adsorbates.

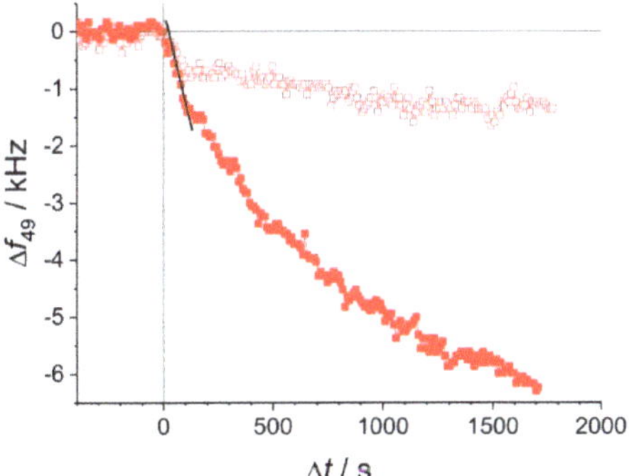

Figure 2. A typical measurement outcome: time course of frequency shift at the 49th overtone of two hydrophobic quartz crystals, under two EV samples of identical dilution, $D = 0.5$ (3.16× dilution of the stock), without freezing (□), and after a 1 h exposure to −30 °C followed by 1 h thermal equilibration at lab temperature (■). The slope of the thick black fitted line represents the maximum rate of frequency shift, $(\mathrm{d}f/\mathrm{d}t)_{\max}$, for the second case.

Interestingly, the duration of frozen storage was found to affect the EMPAS signal in a broad REV concentration range spanning three orders of magnitude. Figure 3A,B illustrate the qualitative changes observable as a result of storage temperature dependence. The maximum rate of frequency change, $r_{\max}$, defined in Equation (1), has been used to express the adsorption rate of the sample components.

$$r_{\max} \overset{\mathrm{def}}{=} -(\mathrm{d}f/\mathrm{d}t)_{\max} \tag{1}$$

The increase of r_{max} with EV concentration is not surprising, as r_{max} is expected to be a symbatic function of the maximum rate of adsorption, which in turn is expected to increase with increasing mass transport, and therefore, concentration. However, surprising is the markedly different dependence of r_{max} on concentration for the samples stored for different time intervals (1 h or 4 d) at −30 °C. The observation suggests that the structural change phenomenon induced by freezing is not instantaneous, but progresses, even at −30 °C, on a timescale of several days. The study of this effect for additional storage time values with identical REV concentrations lead to the result shown in Figure 3B. r_{max} shows a clear, nonlinear, probably exponential dependence on the storage time at −30 °C. Figure 3B also illustrates an interesting, and not yet fully understood, phenomenon about the reversibility of long-term storage effect: namely, the storage time dependent adsorption rate increase can be turned back to a lower value by storing the sample in a thawed state for 4 days. This result might indicate a slow, partial re-aggregation of the REV component molecules after thawing, decreasing the concentration, and re-increasing the size (hence decreasing the diffusion coefficient) of the available adsorbates.

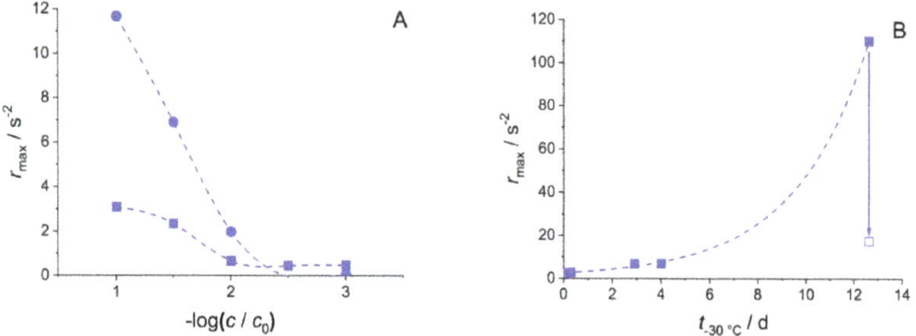

Figure 3. (**A**) r_{max} at the 49th overtone on hydrophilic quartz crystals for different dilutions of the REV stock solution, after a 60 min (■) and 4 days (●) exposure, respectively, to −30 °C. (**B**) Exponential dependence of r_{max} on the storage time at −30 °C of REV $D = 1.5$ solutions (i.e., 31.6× dilution of the original stock), measured at the 49th overtone on hydrophilic quartz crystals (■). Data point marked with □ was measured after at 12.6 d storage at −30 °C, followed by a 4 d storage at +4 °C.

The effect of freeze–thaw cycles on EV functionality has been investigated in multiple cases, but the results are quite variable. Kim et al. [29] observed that four freeze–thaw cycles disrupted the integrity of exosomes (from murine bone marrow-derived dendritic cells), and consequently, the immunosuppressive ability of exosomes was lost. In an untreated sample the exosome-associated heat shock protein Hsc70 was found to be associated with the exosomes but not in the freeze–thaw treated exosome fractions. Lőrincz et al. have found that the storage of neutrophilic granulocyte-derived EVs at room temperature or 4 °C for a day did not influence the vesicle count, but induced a loss of function. Sample storage at −20 °C resulted in changed light scattering properties, and the functionality was almost completely lost after 28 days. After storage at −80 °C for 1 month, light scattering properties did not change, but the number and size of vesicles changed slightly, and the functionality was partially lost [30]. The concentration of exosomes from Sprague Dawley rat bone marrow mesenchymal stem cells was reported to lower after freezing to −80 °C and subsequent thawing. The functionality was also affected [31].

In our case, the changed sample integrity, i.e., protein desorption from the protein corona of the vesicle, might serve as a feasible explanation for the different EMPAS signals in the detection of frozen and non-frozen REVs. To evaluate this possibility, Flu-SEC chromatograms have been measured in consecutive freeze–thaw cycles (Figure 4). Control REVs are eluted at 1.5 min retention time and no sign of free proteins can be observed. Freeze–thaw cycles result in the appearance of a second peak at 4.1 min retention time, which corresponds to free proteins, i.e., to unbound membrane-associated proteins.

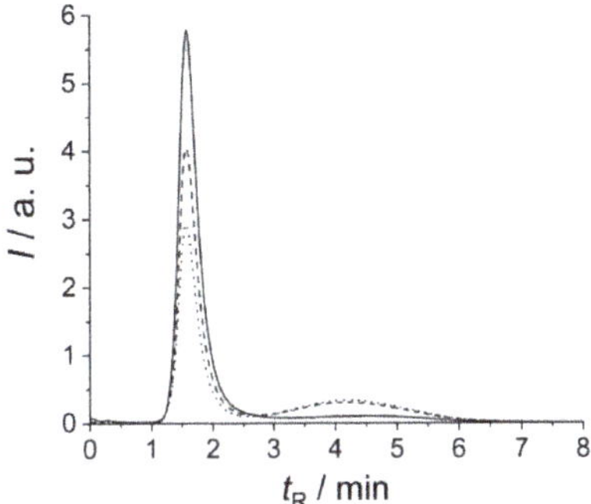

Figure 4. Size exclusion (Flu-SEC) chromatograms (fluorescence intensity *I* versus retention time t_R) of control REVs (solid line), and REVs after one (dashed line) and two (dotted line) freeze–thaw (FT) cycle(s) measured at excitation and emission wavelengths of 280 nm and 330 nm, respectively.

As a sum of observations, we can conclude that intact REVs produce low EMPAS signals, but freezing resulted in disintegration and protein desorption, which appears in a well observable, EV concentration dependent signal. To explain these differences, two effects can be considered: (1) the penetration depth of the acoustic wave, and (2) a large difference in the diffusion coefficients of free protein molecules and EVs. The penetration depth (δ) of the acoustic wave is the distance perpendicular to the surface of the resonating quartz crystal which is reached by the $1/e$-th (~37%) of the energy of the acoustic wave. This distance is a function of the viscosity (η) and density (ρ) of the medium, as well as the ground resonance frequency (f_0) of the crystal and the overtone number (n) at which it is operated. By approximating the viscosity and density of the PBS solution with that of water ($\eta = 10^{-3}$ Pas, $\rho = 10^3$ kg/m^3), and further plugging in $f_0 = 20 \times 10^6$ Hz and $n = 49$, the resulting estimate is $\delta = 18$ nm. Table 1 shows the mean square displacement values of intact vesicles and disintegrated proteins calculated for 180 s time scale, the average duration of stay in the flow cell. By comparing this penetration depth with the mean diameters of EVs (Table 1), it becomes clear that a large frequency shift cannot be obtained if the EVs adsorb as whole, intact globules, not even at high surface coverages; conversely, a large frequency shift is potentially observed if only proteins, or other EV-fragments adsorb, even at lower surface coverages (Figure 5).

$$\delta = \sqrt{\frac{\eta}{\pi \rho n f_0}} \tag{2}$$

Table 1. Comparison of mean square displacement values of intact vesicles and disintegrated proteins calculated for 180 s time scale, the average duration of stay in the flow cell.

Particle	Radius (nm)	Diffusion Coefficient (m^2/s)	Mean Square Displacement (µm)
protein	2.5	9.8×10^{-11}	325.4
vesicle	100	2.5×10^{-12}	51.4

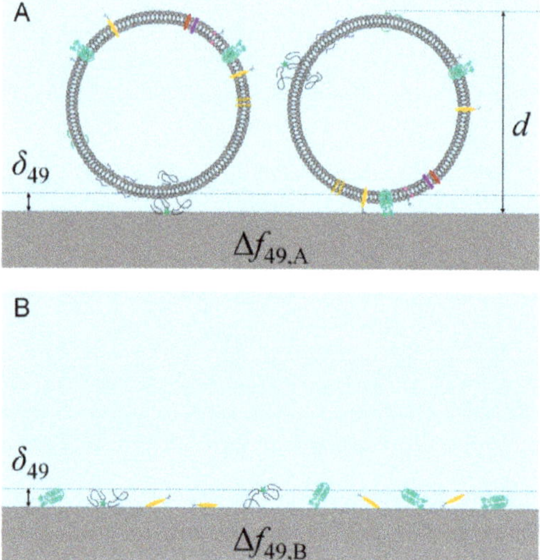

Figure 5. Penetration depth (δ) compared to the size of adsorbed entire extracellular vesicles (EVs) (**A**) and EV-proteins (**B**), respectively, approximately to scale. For maximum packing densities, the frequency shift on the 49th overtone is higher if only proteins adsorb: $\Delta f_B > \Delta f_A$.

3.3. Quantitative Assay Feasibility Study and Sensor Surface Modification for Optimal Detection with EMPAS

Although the results have shown that EMPAS measurements based on non-specific adsorption are insensitive for EVs, the freezing-detached protein fraction of EV suspension can still be a sensitive, quantitatively correlating target to determine. Very importantly, it may even provide a separation-free way to characterize EV isolations in accordance with MISEV 2018 guidelines.

To optimize the adsorption of the protein fraction, measurements using hydrophilic and hydrophobic EMPAS crystals have been conducted. EV dilution series after one freeze–thaw cycle have been measured (Figure 6).

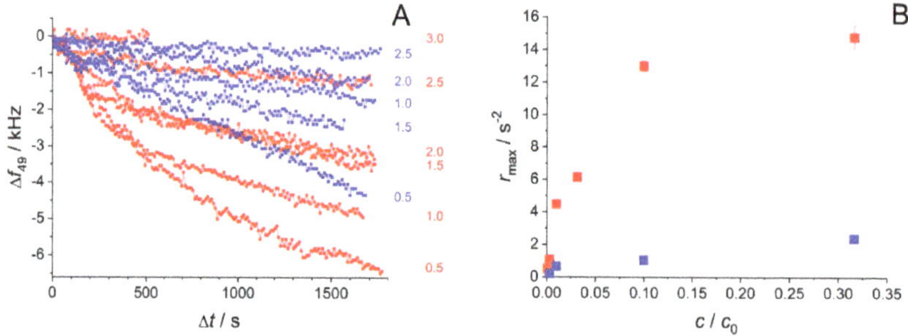

Figure 6. Surface modification effect on the frequency response of electromagnetic piezoelectric acoustic sensor (EMPAS) signal upon adsorption. (**A**) Time course of frequency shift at the 49th overtone on hydrophilic (blue) and hydrophobic (red) quartz crystals, for different dilutions expressed as $-\log(c/c_0)$ of the EV stock solution, after a short exposure (less than 1 day) to $-30\ °C$. (**B**) r_{max} at the 49th overtone on hydrophilic (blue) and hydrophobic (red) quartz crystals, for different dilutions of the REV stock solution, after a short exposure (less than 1 day) to $-30\ °C$.

A comparison of hydrophobic and hydrophilic surfaces shows that for the same EV dilution the EMPAS signal has a greater frequency shift in the case of hydrophobic surfaces, indicating a higher adsorption rate. Although vesicle quantification cannot be directly achieved with the present configuration, the amount of measurable free proteins adsorbed on the crystal correlates to the number of vesicles, and thus the methodology developed may serve as an indirect method for quantification. For this approximation, we use the protein concentration calculated from IR measurements given as $c_0 = 950$ µg/mL. Based on Flu-SEC measurements, we assume that 30% of the total protein is detached, and as the most diluted sample is $-\log(c/c_0) = 2.5 \approx 0.3\%$ dilution, the detection limit is in the range of 0.9 µg/mL of protein. The vesicle concentration in the original sample is $c_{REV} = 5.6 \times 10^{11}$ mL^{-1} according to MPRS measurements, so the detection limit for vesicle concentration in the given conditions is in the range of 10^{-9} mL^{-1}. By considering the average size ($r = 100$ nm) and concentration of REVs measured by MRPS and estimating the protein size with the size of an average human protein, we arrive at a realistic approximation of 0.15 protein/nm^2 of vesicle for the given experimental conditions. It is important to emphasize that the detection limit is valid for the proteins in the EV suspension, hence in the presence of EVs, without separation.

4. Conclusions

The adsorption properties of well-characterized red blood cell–derived extracellular vesicles have been investigated with a high sensitivity electromagnetic piezoelectric acoustic sensor. Maximum rates of frequency shifts of the oscillating sensor have been compared to represent differences in adsorption capacities. Freeze–thaw pretreatment of the samples has largely affected the measurable signal, indicating that the disrupted integrity of the vesicles, verified with chromatography, results in well-detectable substances. Hydrophobic surfaces proved to adsorb faster sample content than hydrophilic surfaces did. The phenomenon is well known from the biofouling field: in aqueous environment, low surface energy (hydrophobic) surfaces tend to adsorb faster than high surface energy (hydrophilic) surfaces, especially when the adsorbates are macromolecules, such as (bio)polymers (e.g., proteins). The most plausible explanation is based on the total entropy gain of the system upon releasing the hydrate shell water molecules during the adsorption of the macromolecule. The maximum rates of frequency shifts increased with increasing initial vesicle concentration. As such, this method provides the possibility of indirect vesicle quantification even without separation steps applied. The assay can also be considered as a follow-up technique in sample storage standardization processes.

Author Contributions: Conceptualization, L.R., Z.V., Z.K. and M.T.; methodology, L.R., Z.V., J.M., Z.K. and M.T.; validation, L.R., Z.V. and J.M.; formal analysis, L.R., Z.V. and J.M.; resources, Z.K. and M.T.; data curation, L.R., Z.V. and J.M.; writing—original draft preparation, L.R., Z.V., J.M. and Z.K.; writing—review and editing, L.R., Z.V., J.M., Z.K. and M.T.; supervision, M.T. All authors have read and agreed to the published version of the manuscript.

Funding: This work was supported by the European Commission under grant agreement number 690898/H2020-MSCA-RISE-2015, the National Competitiveness and Excellence Program Hungary, (NVKP_16-1-2016-0007) and BIONANO GINOP-2.3.2-15-2016-00017 projects. Z.V. was supported by the János Bolyai Research Scholarship of the Hungarian Academy of Sciences. The funding by the National Research, Development and Innovation Office NKFIH under grant number K131594 is also acknowledged (JM).

Conflicts of Interest: The authors declare no conflict of interest.

References

1. Properzi, F.; Logozzi, M.; Fais, S. Exosomes: The future of biomarkers in medicine. *Biomark. Med.* **2013**, *7*, 769–778. [CrossRef] [PubMed]
2. Tang, Z.; Li, D.; Hou, S.; Zhu, X. The cancer exosomes: Clinical implications, applications and challenges. *Int. J. Cancer* **2020**, *146*, 2946–2959. [CrossRef] [PubMed]
3. Théry, C.; Witwer, K.W.; Aikawa, E.; Alcaraz, M.J.; Anderson, J.D.; Andriantsitohaina, R.; Antoniou, A.; Arab, T.; Archer, F.; Atkin-Smith, G.K.; et al. Minimal information for studies of extracellular vesicles 2018 (MISEV2018): A position statement of the International Society for Extracellular Vesicles and update of the MISEV2014 guidelines. *J. Extracell. Vesicles* **2018**, 1535750. [CrossRef] [PubMed]

4. van der Pol, E.; de Rond, L.; Coumans, F.A.W.; Gool, E.L.; Böing, A.N.; Sturk, A.; Nieuwland, R.; van Leeuwen, T.G. Absolute sizing and label-free identification of extracellular vesicles by flow cytometry. *Nanomed. Nanotechnol. Biol. Med.* **2018**, *14*, 801–810. [CrossRef]

5. Armitage, J.D.; Tan, D.B.A.; Cha, L.; Clark, M.; Gray, E.S.; Fuller, K.A.; Moodley, Y.P. A standardised protocol for the evaluation of small extracellular vesicles in plasma by imaging flow cytometry. *J. Immunol. Methods* **2019**, *468*, 61–66. [CrossRef]

6. Gardiner, C.; Ferreira, Y.J.; Dragovic, R.A.; Redman, C.W.G.; Sargent, I.L. Extracellular vesicle sizing and enumeration by nanoparticle tracking analysis. *J. Extracell. Vesicles* **2013**, *2*, 19671. [CrossRef]

7. Desgeorges, A.; Hollerweger, J.; Lassacher, T.; Rohde, E.; Helmbrecht, C.; Gimona, M. Differential fluorescence nanoparticle tracking analysis for enumeration of the extracellular vesicle content in mixed particulate solutions. *Methods* **2020**, *177*, 67–73. [CrossRef]

8. Coumans, F.A.W.; van der Pol, E.; Böing, A.N.; Hajji, N.; Sturk, G.; van Leeuwen, T.G.; Nieuwland, R. Reproducible extracellular vesicle size and concentration determination with tunable resistive pulse sensing. *J. Extracell. Vesicles* **2014**, *3*, 25922. [CrossRef]

9. Maas, S.L.N.; De Vrij, J.; Broekman, M.L.D. Quantification and size-profiling of extracellular vesicles using tunable resistive pulse sensing. *J. Vis. Exp.* **2014**, *92*, e51623. [CrossRef]

10. Vogel, R.; Coumans, F.A.W.; Maltesen, R.G.; Böing, A.N.; Bonnington, K.E.; Broekman, M.L.; Broom, M.F.; Buzás, E.I.; Christiansen, G.; Hajji, N.; et al. A standardized method to determine the concentration of extracellular vesicles using tunable resistive pulse sensing. *J. Extracell. Vesicles* **2016**, *5*, 31242. [CrossRef]

11. Wang, Q.; Zou, L.; Yang, X.; Liu, X.; Nie, W.; Zheng, Y.; Cheng, Q.; Wang, K. Direct quantification of cancerous exosomes via surface plasmon resonance with dual gold nanoparticle-assisted signal amplification. *Biosens. Bioelectron.* **2019**, *135*, 129–136. [CrossRef] [PubMed]

12. Rupert, D.L.M.; Lässer, C.; Eldh, M.; Block, S.; Zhdanov, V.P.; Lotvall, J.O.; Bally, M.; Höök, F. Determination of exosome concentration in solution using surface plasmon resonance spectroscopy. *Anal. Chem.* **2014**, *86*, 5929–5936. [CrossRef] [PubMed]

13. Thakur, A.; Qiu, G.; NG, S.P.; Guan, J.; Yue, J.; Lee, Y.; Wu, C.M.L. Direct detection of two different tumor-derived extracellular vesicles by SAM-AuNIs LSPR biosensor. *Biosens. Bioelectron.* **2017**, *94*, 400–407. [CrossRef] [PubMed]

14. Chang, Y.J.; Yang, W.T.; Wu, J.C. Isolation and detection of exosomes via AAO membrane and QCM measurement. *Microelectron. Eng.* **2019**, *216*, 111094. [CrossRef]

15. Thompson, M.; Ballantyne, S.M.; Cheran, L.E.; Stevenson, A.C.; Lowe, C.R. Electromagnetic excitation of high frequency acoustic waves and detection in the liquid phase. *Analyst* **2003**, *128*, 1048–1055. [CrossRef]

16. Ballantyne, S.M.; Thompson, M. Superior analytical sensitivity of electromagnetic excitation compared to contact electrode instigation of transverse acoustic waves. *Analyst* **2004**, *129*, 219–224. [CrossRef]

17. Sheikh, S.; Blaszykowski, C.; Thompson, M. Label-free detection of HIV-2 antibodies in serum with an ultra-high frequency acoustic wave sensor. *Talanta* **2011**, *85*, 816–819. [CrossRef]

18. Sheikh, S.; Blaszykowski, C.; Romaschin, A.; Thompson, M. Endotoxin detection in full blood plasma in a theranostic approach to combat sepsis. *RSC Adv.* **2016**, *6*, 38037–38041. [CrossRef]

19. Neves, M.A.D.; Blaszykowski, C.; Bokhari, S.; Thompson, M. Ultra-high frequency piezoelectric aptasensor for the label-free detection of cocaine. *Biosens. Bioelectron.* **2015**, *72*, 383–392. [CrossRef]

20. Crivianu-Gaita, V.; Aamer, M.; Posaratnanathan, R.T.; Romaschin, A.; Thompson, M. Acoustic wave biosensor for the detection of the breast and prostate cancer metastasis biomarker protein PTHrP. *Biosens. Bioelectron.* **2016**, *78*, 92–99. [CrossRef]

21. Sheikh, S.; Yang, D.Y.; Blaszykowski, C.; Thompson, M. Single ether group in a glycol-based ultra-thin layer prevents surface fouling from undiluted serum. *Chem. Commun.* **2012**, *48*, 1305–1307. [CrossRef] [PubMed]

22. De Los Santos Pereira, A.; Sheikh, S.; Blaszykowski, C.; Pop-Georgievski, O.; Fedorov, K.; Thompson, M.; Rodriguez-Emmenegger, C. Antifouling Polymer Brushes Displaying Antithrombogenic Surface Properties. *Biomacromolecules* **2016**, *17*, 1179–1185. [CrossRef] [PubMed]

23. Spagnolo, S.; de la Franier, B.; Hianik, T.; Thompson, M. Surface probe linker with tandem anti-fouling properties for application in biosensor technology. *Biosensors* **2020**, *10*, 20. [CrossRef] [PubMed]

24. Románszki, L.; Tatarko, M.; Jiao, M.; Keresztes, Z.; Hianik, T.; Thompson, M. Casein probe–based fast plasmin determination in the picomolar range by an ultra-high frequency acoustic wave biosensor. *Sens. Actuators B Chem.* **2018**, *275*, 206–214. [CrossRef]

25. Kitka, D.; Mihály, J.; Fraikin, J.L.; Beke-Somfai, T.; Varga, Z. Detection and phenotyping of extracellular vesicles by size exclusion chromatography coupled with on-line fluorescence detection. *Sci. Rep.* **2019**, *9*, 19868. [CrossRef]

26. Szentirmai, V.; Wacha, A.; Németh, C.; Kitka, D.; Rácz, A.; Héberger, K.; Mihály, J.; Varga, Z. Reagent-free total protein quantification of intact extracellular vesicles by attenuated total reflection Fourier transform infrared (ATR-FTIR) spectroscopy. *Anal. Bioanal. Chem.* **2020**, *412*, 4619–4628. [CrossRef]

27. Visnovitz, T.; Osteikoetxea, X.; Sódar, B.W.; Mihály, J.; Lőrincz, P.; Vukman, K.V.; Tóth, E.Á.; Koncz, A.; Székács, I.; Horváth, R.; et al. An improved 96 well plate format lipid quantification assay for standardisation of experiments with extracellular vesicles. *J. Extracell. Vesicles* **2019**, *8*, 1565263. [CrossRef]

28. Mihály, J.; Deák, R.; Szigyártó, I.C.; Bóta, A.; Beke-Somfai, T.; Varga, Z. Characterization of extracellular vesicles by IR spectroscopy: Fast and simple classification based on amide and CH stretching vibrations. *Biochim. Biophys. Acta Biomembr.* **2017**, *1859*, 459–466. [CrossRef]

29. Kim, S.-H.; Lechman, E.R.; Bianco, N.; Menon, R.; Keravala, A.; Nash, J.; Mi, Z.; Watkins, S.C.; Gambotto, A.; Robbins, P.D. Exosomes Derived from IL-10-Treated Dendritic Cells Can Suppress Inflammation and Collagen-Induced Arthritis. *J. Immunol.* **2005**, *174*, 6440–6448. [CrossRef]

30. Lorincz, Á.M.; Timár, C.I.; Marosvári, K.A.; Veres, D.S.; Otrokocsi, L.; Kittel, Á.; Ligeti, E. Effect of storage on physical and functional properties of extracellular vesicles derived from neutrophilic granulocytes. *J. Extracell. Vesicles* **2014**, *3*, 25465. [CrossRef]

31. Teng, X.; Chen, L.; Chen, W.; Yang, J.; Yang, Z.; Shen, Z. Mesenchymal stem cell-derived exosomes improve the microenvironment of infarcted myocardium contributing to angiogenesis and anti-inflammation. *Cell. Physiol. Biochem.* **2015**, *37*, 2415–2424. [CrossRef] [PubMed]

Publisher's Note: MDPI stays neutral with regard to jurisdictional claims in published maps and institutional affiliations.

 *biosensors*

Article

Detecting Vasodilation as Potential Diagnostic Biomarker in Breast Cancer Using Deep Learning-Driven Thermomics

Bardia Yousefi [1,*], Hamed Akbari [2] and Xavier P.V. Maldague [1,*]

[1]   Department of Electrical and Computer Engineering, Laval University, Quebec City, QC G1V 0A6, Canada
[2]   Department of Radiology, University of Pennsylvania, Philadelphia, PA 19104, USA; AkbariHA@upenn.edu
*   Correspondence: Bardia.Yousefi.1@ulaval.ca (B.Y.); Xavier.Maldague@gel.ulaval.ca (X.P.V.M.)

Received: 23 September 2020; Accepted: 30 October 2020; Published: 31 October 2020

Abstract: Breast cancer is the most common cancer in women. Early diagnosis improves outcome and survival, which is the cornerstone of breast cancer treatment. Thermography has been utilized as a complementary diagnostic technique in breast cancer detection. Artificial intelligence (AI) has the capacity to capture and analyze the entire concealed information in thermography. In this study, we propose a method to potentially detect the immunohistochemical response to breast cancer by finding thermal heterogeneous patterns in the targeted area. In this study for breast cancer screening 208 subjects participated and normal and abnormal (diagnosed by mammography or clinical diagnosis) conditions were analyzed. High-dimensional deep thermomic features were extracted from the ResNet-50 pre-trained model from low-rank thermal matrix approximation using sparse principal component analysis. Then, a sparse deep autoencoder designed and trained for such data decreases the dimensionality to 16 latent space thermomic features. A random forest model was used to classify the participants. The proposed method preserves thermal heterogeneity, which leads to successful classification between normal and abnormal subjects with an accuracy of 78.16% (73.3–81.07%). By non-invasively capturing a thermal map of the entire tumor, the proposed method can assist in screening and diagnosing this malignancy. These thermal signatures may preoperatively stratify the patients for personalized treatment planning and potentially monitor the patients during treatment.

Keywords: vasodilator activity; breast cancer screening; imaging biomarker; deep sparse autoencoder; dimensionality reduction; deep-learning features

---

1. Introduction

Breast cancers caused an estimated 41,760 deaths out of 606,808 overall deaths for females and 500 deaths for males, while the estimated new cases were 271,270 deaths for both genders (268,600 women and 2670 men) in the United States in 2019. This evidence shows that despite considerable advancement in breast cancer screening and treatment, breast cancer is still the second cause of cancer death among women [1]. Clinical breast exam (CBE), magnetic resonance imaging (MRI), mammography, and ultrasound are widely used for the diagnosis of breast cancer. Among them, CBE and mammography are considered the most common breast screening tools [1]. This study proposes machine learning techniques and analyses using infrared thermography as a new technique for breast cancer screening. We hypothesized that thermal heterogeneity may associate with angiogenesis, nitric oxide vasodilatory phenomena, inflammation, and estrogen caused by cancer symptoms.

Mammography has been the gold standard for diagnosing breast cancer since the early 1960s despite numerous studies indicating the variability of this imaging modality is affected by breast density, age, type of problem, and family history [2–5]. Mammography showed weakness in being used for breast cancer screening for women with dense breasts, hormone replacement therapy, and fibrocystic breasts [6–9]. Research showed the detection rate of mammography considerably diminishes by increasing breast density among the patients [7]. The studies show that age has a reverse association with breast cancer detection using mammography [10]. Moreover, findings show that during mammography the pressure on the breast tumors is adequate to rupture the encapsulated tumors (depends on the location of the tumor) and possibly circulate malignant cells in the bloodstream [11].

The risk of radiation using mammography is one of the important disadvantages of using this imaging modality. Younger women are more susceptible to the risk of radiation-induced breast cancer than older women due to their undifferentiated cells being prone to influence by ionizing radiation [5,10–15].

Family history of breast cancer and/or the *BRCA1/2* gene mutations are other factors that this gene mutation may be the result of radiation effect, which might be induced indirect effects to damage DNA by producing reactive oxygen species (ROS) from the cell's water molecules [16,17]. Some studies showed that the ionizing radiation by mammography might be more dangerous for patients with such mutations [16–19].

For the dense breast, ultrasound can be an adjunctive tool used with mammography screening for detecting the abnormality [2]. However, it shows a dependency on tumor size, palpability, breast density, tools' quality, physician's expertise who performs the procedure, and interpreting the image [2,5,20–22]. Magnetic resonance imaging (MRI) is also an alternative imaging modality, which can identify early breast cancer in the place where conventional imaging fails to detect the abnormalities [23,24]. Ahen et al. (2014) concluded that the high costs and low specificity of MRI limits the popularity of MRI for annual screening for high-risk women [25]. CBE is considered as a great alternative conducted by the clinician and can help to detect at least 50% of asymptomatic breast cancers but has not been used alone [26,27].

2. Thermography and Biological Rationale as an Alternative Imaging Modality

Infrared thermography is used as an additional cost-effective alternative for breast cancer screening as a non-invasive procedure that does not pressurize the breast tissue nor expose the body to ionizing radiation. The skin emission is about 0.98, which is close to the emission of the blackbody. Thermal radiation emitted from the body has a wavelength of 8–10 μm bandwidth, which can be captured by the infrared camera [28–30]. Due to the relatively lower sensitivity of thermographic screening, it usually adds to other diagnosing methods, mainly with CBE to increase the overall diagnostic accuracy [31,32].

Blood circulation is the main contributor to heat transfer in the body. Vascularity is also considered an important parameter for heat transfer [33]. Evidence supports different thermal conductivity between normal tissues in breasts and cancerous lesion thermal profile discrepancy [34], also abnormal skin temperature manifestation is an indicator of pathological changes explained by metabolic activity associated with the tumor such as angiogenesis, nitric oxide, inflammation, and estrogen [5,33,35]. Changing in the endocrine due to the presence of tumors alters the thermal profile by changing the vascularization of the tissues to deliver oxygen and nutrients to tumors [36], in the process of pathologic angiogenesis. In such chaotic and pathological processes of angiogenesis, smooth muscle cells receiving abnormal vasoconstrict blood vessels in the area [36]. Several studies proved the value of infrared thermography on detecting hypervascularity and hyperthermia on non-palpable breast cancer [36–38].

Tumor angiogenesis and metastatic behavior are biomarkers of breast carcinoma and c-Met pathway activation, which are used also for tumor progression. Also, there is an association of c-Met and downstream signaling pathways with angiogenesis that can be assessed by microvessel density (MVD) [39]. Infrared was used to assess the existence of MVD, which is associated with endothelial cell (CD34) marker and downstream signaling pathways (angiogenesis, RAS-MAPK, and PI3K-AKT) [39]. In another study, there is a relationship between MVD and tumor-associated macrophages (TAMs) and vascular endothelial growth factor (VEGF) expression [23,40,41]. In addition, white blood cells produce nitric oxide as a defense mechanism against cancerous cells that is a vasodilatory substance [31,37]. Nitric oxide performs as a vasodilator in cancerous tissues in the breasts to enhance oxygen and nutrient delivery, which increases the local temperature in the area [42–44]. In general, cancer (including breast cancer) is influenced by different cellular factors including reactive oxygen and nitrogen species (RONS) and, as reported, hypoxic condition elevated RONS production [44–47]. Nitric oxide has reactive diatomic free radical plays a role in promoting and inhibiting cancer [48].

DNA damage occurs once nitric oxide reacts to form other reactive metabolites as well as nitrite, peroxynitrite, nitrate, or S-nitroso-thiols, which provoke genotoxic effects [48,49]. Reactive nitrogen species (RNS) exposure causes post-translational changes, which leads to different interactions by other cellular targets and causes diverse locally dependent concentration effects [49,50]. In the late stages of breast cancer also, c-Met promotes metastases by having vascular reprogramming and inflammatory cytokine upregulation [51], inflammation-related cytokine tumor necrosis factor-alpha (TNF-α) in tumor invasion [52]. This often happens due to a long period of remission before the diagnosis of breast cancer [53], which can be detected faster using the combination of CBE and infrared screening. The presence of inflammation is another mechanism of local heat generation. Cancer causes a vasodilatory response, due to the inflammatory cell involvement' which increases temperature [42]. Estrogen also facilitates vasodilation by locally enhancing nitric oxide production. Imbalanced estrogen could change the vasodilatory effects of the tissue resulting in thermal variations [54]. Evidence shows that ecto-5′- nucleotidase (eN) is negatively controlled by estrogen receptor-α (ERα). This suggests that eN expression and its adenosine generation associate with breast cancer progression. eN expression in estrogen receptor-negative cells considered to be an aggressive breast cancer biomarker [55].

Such process metabolic heat generation investigated for normal and cancerous breast tissues and its rates reached 20K W/m^3 and the range between 100K–1200K W/m^3 for two types of tissues, respectively [56]. Despite United States Food and Drug Administration (FDA) approval for using infrared thermography, it can be used as an adjunct screening modality along with MRI, mammography, and ultrasound [57,58]. A visual summary of the factors influencing the heterogeneity in thermal imaging are presented in Figure 1.

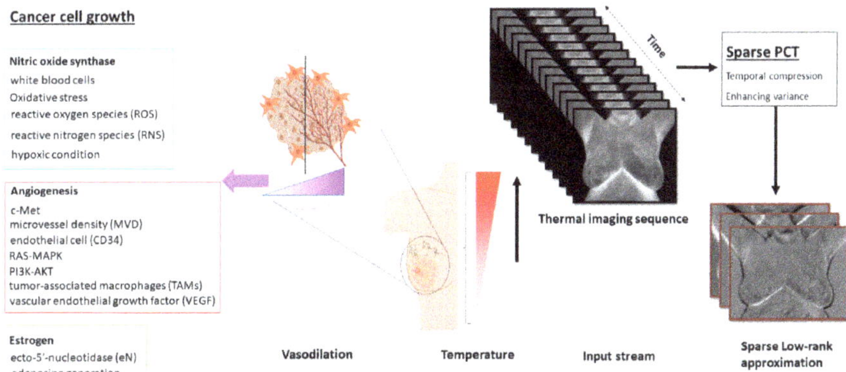

Figure 1. The block diagram of the biological connection to the response of infrared thermography as a fast step with other methods such as clinical breast exam (CBE) in breast cancer screening and cancer presence in the breast area are shown.

In this study, we propose a method to use high-dimensional deep-learning features to track the vasodilator activities in the breast area as a potential biomarker in detecting breast cancer. The contributions of the paper are as follows:

- The sparse principal component analysis in the thermography (Sparse PCT) is used to compress the input thermal sequence and capture high temporal variance across the acquisitions. This leads to capture thermal heterogeneity patterns in three first initial bases, called *avatars*, which concatenate in three different channels similar to a red, green and blue (RGB) image as the input for our pretrained model.
- Deep thermal features, called *deep-thermomics* inspired by radiomics, are extracted to measure thermal heterogeneity in breast cancer screening using infrared thermography.
- The proposed approach tackles the problem of the *curse of dimensionality* in deep-thermomics using a sparse deep autoencoder without using traditional human-engineered feature selection methods.
- The multivariate models trained and validated using the obtained descriptors successfully classify between symptomatic and non-symptomatic subjects. We also provided a comparative analysis using a non-sparse PCT.
- This study shows the association between thermal heterogeneous patterns and potential vasodilation in the breast area, as a new potential imaging biomarker.

The rest of the paper is organized as follows. In the next section, thermal transfer in passive thermography is summarized. In Section 3, the methodology of the approach will be briefly described by applying sparse PCT analysis for thermography and pre-trained ResNet-50 deep neural networks. The experimental results are presented in Section 5, and the discussion is in Section 6. The conclusions are summarized in Section 7.

3. Thermal Transfer in Thermography

A thermal camera captures the spatial heterogeneity of temperature on the targeted region of interest (ROI) over time. This heat transient can be through active or passive thermography techniques. In general, the thermal transfer/heat conduction equation of a specimen can be summarized by the following equation:

$$\rho C_p \frac{\partial T}{\partial t} = k \frac{\partial^2 T}{\partial t^2} + \dot{q} \tag{1}$$

where $T = T(x, y, z)$ is a temperature field, k is thermal conductivity constant from the material (W/m.K). ρ is the density $\left(\text{kg/m}^3\right)$, C_p is specific heat (J/kg.K), $\dot{q}(x, y, z, t)$ is the internal heat generation function per unit volume, in the passive thermography.

Applying infrared thermography on biological organs and tissues, as a complex structure, composed of fat, blood vessels, parenchymal tissues, and nerves with some uncertainty for the rate of blood perfusion and metabolic activity. Pennes' bioheat equation [59] provides accurate thermal computations and states as follows:

$$\rho_t c_t \left(\frac{\partial T_t}{\partial t} \right) = \nabla.(k_t \nabla T_t) + \omega_b c_b (T_a - T_t) + q_m \tag{2}$$

where ω_b represents the flow rate of blood, q_m is the metabolic rate (heat generation), and b, and a in $\omega_b c_b (T_a - T)$ the additive term stands for blood, and arteries (in targeted tissue), respectively.

4. Methodology

Infrared thermography records thermal heterogeneity in the subdermal area of the breast in temporal order. To capture such effect by abstracting such patterns, low-rank matrix approximation is used to maximize the variance across thermal acquisition time.

4.1. Low-Rank Approximation of Thermal Stream

Low-rank matrix approximation is commonly used in thermography [60–68], due to capturing thermal variations across the temporal order in the sequence. This leads to detecting thermal patterns on the subsurface of specimens. Such analyses capture thermal heterogeneity in the skin area for breast cancer screening patients. Principal component analysis (PCA), called PCT for thermography [60], through singular value decomposition (SVD) used for decomposing the input matrix (heat matrix) X, which is $p \times n$, where n is the vectorized thermal image (breast screening) in every sequence and p corresponds to the number of observations and decomposes to:

$$X = U\Gamma V^T \tag{3}$$

where U is the $p \times n$ matrix $(p > n)$ and Γ is a diagonal matrix with a dimension of $n \times n$ and either zero or positive elements. V^T denotes the transpose of the $n \times n$ matrix. The method captures the spatiotemporal variance by selecting the bases correspond to 80% of variance from the eigenvector matrix. Matrix U represents the bases of the input matrix.

The PCT is a linear transformation technique that decomposes the input zero-mean data matrix into the bases and coefficient matrix. To find the optimal solution for such transformation, ℓ_2 and ℓ_1 penalty terms with regularization parameters were added to the PCT, which led to Sparse PCT [61,62] and increased the performance of such a technique, particularly when encountering additive noise.

Such modifications in Sparse PCT not only turned PCT into a nonlinear transformation but following the same maximization of the variance among the bases. If the empirical covariance matrix of $X_{p \times n}$ is presented by $X^T X$, Sparse PCT is the maximization of variance in the direction of vector $v \in \mathbb{R}^p$ for $1 \geq k \geq p$.

$$max \ v^T \Sigma v$$
$$such \ that \ \| \ v \ \|_2 = 1 \ , \ \| \ v \ \|_0 \geq k \tag{4}$$

Let v be a variance of the input matrix and $\| \ v \ \|_0$ be ℓ_0 norm of v, which is the non-zero components. This is an NP-hard (non-deterministic polynomial-time hard) problem and Zou et al. (2006) and elastic net algorithms used to solve this optimization [69]. Sparse PCT showed considerable performance in thermography to find a low-rank approximation of input thermal images. Here, we applied sparse PCT to preserve thermal heterogeneity in the subsurface of skin as a potential biomarker leading to early diagnosis of breast cancer.

Let I is thermal imaging stream taken from the participants such that $I \in \mathbb{R}^{n \times m \times \tau}$. If x is a vectorized matrix named heat matrix made by stacking vectorized infrared images, $x \in \mathbb{R}^{n.m \times k}$. $B = \{\beta_1, \beta_2, \ldots, \beta_\tau\}$ denotes a set of bases obtained by sparse PCT. Each β cropped to a squared matrix focusing more on the ROI as an input to the ResNet-50 (spatial dimension of 224×224). Using $k = 3$ corresponding to three predominant low-rank matrix approximation, we capture dynamic variations on thermal images in the ROI during τ time.

4.2. Deep Thermomics

Deep neural networks and particularly convolutional neural networks (CNN) are widely used by researchers in various fields with diverse applications, comprising image processing, and particularly medical imaging. CNN is a group of connected deep neural networks that uses a variation of multi-layer perceptron with many hidden layers [70]. The hidden layers of CNN normally consist of convolutional (cross-correlation) layers (filtering), pooling layers, rectifier layer (ReLu), fully connected layers, and normalization layers [71,72]. Several adaptive filters (as kernels) with small receptive fields layers makes CNN different from other similar deep neural networks. Because of such filtering in the input layers using a 2-dimensional dot product between the filter entries and the input data, the model extracts some features with higher sensitivity in spatial positions of input. This increases the applications of CNN-based networks in a variety of applications with the focus of imaging. Some of these networks are already trained for specific imaging datasets and used as a pre-trained network to perform classification or recognition.

After the success of the AlexNet [72] in image processing at the LSVRC2012, deep residual network (ResNet) [73] was perhaps the most innovative research in the computer vision and deep learning research community. ResNet provides the ability of a trainable network with many layers while holding a compelling performance.

The state-of-the-art CNN architectures are going deeper such as the very deep CNN for large-scale visual recognition (VGG) network [74], GoogleNet (also codenamed Inception-v1) [75] that have 19 and 22 layers respectively. ResNet tackle the vanishing gradient issue by introducing an "identity shortcut connection" that skips one or more layers [73], which does not degrade the network performance, since it simply stacks identity mappings in every layer. The pre-activation variant of residual block [76] increases the popularity of ResNet with an excessive number of hidden layers in the computer vision and medical image processing.

A deep learning method has been employed to non-invasively detect chemically treated collagenous tissue nonlinear anisotropic stress-strain responses in the microscopic images [77]. VGG16 is used for the prognosis of glioblastoma and as a radiographic biomarker for noninvasive categorization between true progression and pseudo-progression in these patients [78]. The initial application of the deep features in infrared analyses has been presented for finding defective patterns in the specimens using spectral difference among various areas of specimens [79]. Using traditional dimensionality reduction or feature selection is not a substantial way due to the low-level status of hidden weights in this model, which might be perceived as collinearity among features. In infrared breast cancer screening methods, a comparative analysis on AlexNet, GoogLeNet, ResNet-18, VGG-16, and VGG-19 for 88 patients using a pre-trained model resulted in discrimination between normal and pathologic patients [80]. ResNet50 was applied to extract features from histopathological images and followed by autoencoder, K-means clustering to choose discriminative patches using PCA to diagnose breast cancer [81]. A CNN approach tackled the same dataset using ResNet34 and ResNet50 and achieved a significant performance on detection of breast cancer in blind validation, and used the entire thermal sequences as the input of their system [82]. A cohort of 57 cases used applying a new configuration of CNN showed promising accuracy, while they outperformed ResNet50, SeResNet50 and Inception models [83]. Similarly, CNN used with additional algorithms such as with Bayes algorithm [84] or support vector machine (SVM) [85] to conduct diagnosis assessments.

In this paper, a pre-trained residual deep convolutional network for large-scale image recognition (ResNet-50) [73,86] was used. ResNet-50 is used as a hybrid feature generator (deep features). The method uses low-rank matrix approximation of thermal sequences as a sparse representation of the whole set, called *avatar*. Three first bases make three channels representing the entire thermal set as an input to the ResNet-50 model and extracted deep-thermomic features, 2048 size vector, as output.

Since the input of the ResNet-50 model requires an RGB squared image, we leverage this property to embed three first bases obtained by sparse PCT as three channels of the input image, showed by ψ, where $\psi_{RGB} \longrightarrow \psi_{\beta_1\beta_2\beta_3}$ (see Figure 2). Applying feed-forward convolution in neural network lookalikes of multiple-internal-functions gives:

$$\mathcal{F}(\psi) = \mathcal{F}_L(\ldots\mathcal{F}_2(\mathcal{F}_1(\psi;\{W_1\});\{W_2\})\ldots;\{W_L\}). \qquad \mathcal{F}: \mathbb{R}^{224\times224\times3} \qquad (5)$$

Let $\mathcal{F}_i$ represents the residual mapping to be learned and $\{W_i\}$ denotes weights in each layer. The regular linear convolution involves a filter bank where the output also contains the input dimensional property. The last layer contains a vector by the size of 2048 and links to three channels low-rank representation of infrared stream. This gives 2048 low-level features from the image used as input to the deep learning-based dimensionality reduction model.

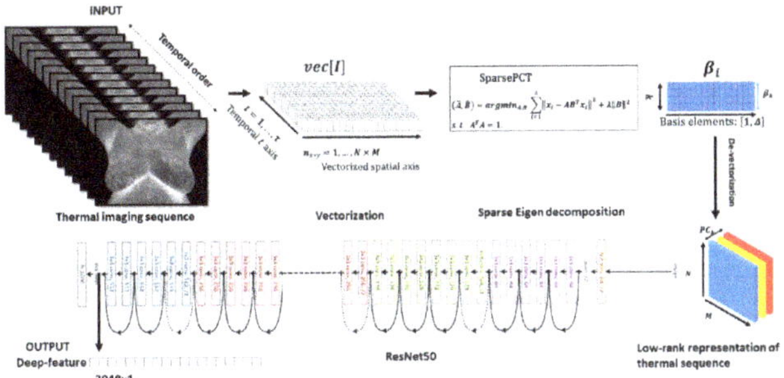

Figure 2. Workflow of the proposed approach in temporal compression and extraction of low-rank matrix approximation and generating the deep thermomics using residual network (ResNet-50) is presented.

4.3. Sparse Autoencoder for Dimensionality Reduction for Deep Thermomics

Radiomics, high-throughput features, refer to sub-visual/quantitative feature extraction and consider a vital part of medical image analysis and radiology which strive to exploit the amount of quantitative minable features extract from imaging data [87]. Every feature contains a distinct phenotype of the tumor which may have diagnosis/prognostic power and adjunctive clinical importance across the different diseases. In oncology, identified features from radiology or any other imaging data facilitate prediction, diagnosis, and prognosis associated with cancer disease to monitor the response, like survival, as a progression criterion of disease and treatment response. Pretrained deep neural networks provide high-dimensional features as an opportunity to gain information on tumor area and its environment that is not otherwise available to the radiologist.

Having sufficed features from the medical images leverage better diagnostic/prognostic decisions whereas high-dimensional feature space can impede computation and enervates the performance of feature selection known as the *curse of dimensionality* problem. This creates a wrong outcome of the model due to overfitting the decision-making unit. Traditional feature selection might not the best solution for such an issue because the low-level informative features provided by the network might translate as collinearity among the descriptors and lead to the elimination of valuable information. Here, we propose an autoencoder trained specifically for such high-dimensional throughput features to reduce the dimensionality hierarchically to the lower dimension. Autoencoders are data compression algorithms that make of hierarchical compression and decompression units cascade to each other. They are data-specific, automatically learn from training input data instead of being obtained by human interference, and lossy. Autoencoders can compress data like what they have been trained on and cannot be generalized for other dissimilar data, while they are different from lossless arithmetic compression [88]. An autoencoder contains encoding and decoding parametric functions with a measure of distance, or "loss" function, between the compressed representation of the input data and the final decompressed representation. The parameters of the encoding/decoding functions can be optimized by using stochastic gradient descent to minimize the reconstruction loss.

Autoencoder architecture. Several dense layers with different sizes were employed to reduce the dimensionality from 2048 to 16 compressed descriptors, in the *latent space*. Eight dense (8D) layers, including 4 dense layers in each of encoder and decoder, were used. The intermediate representation of feature dimensionality was varied from the size of 1024, 256, and then 64, to latent space with size 16. Each layer has a ReLu activation function and the last layer has a Sigmoid activation function, with sparse constraints in the initial layer (Figure 3). The network trained for the batch size of 128, with a total of 3000 iterations, with an unfixed learning rate in the Adam optimization algorithm.

Let $x \in \mathbb{R}^F$ considers as the first mapped input, where $F = 2048$, to the latent space with $h = f_e(x) = a_e(Wx + b_e)$ is the hidden representation of the input vector, a_e is the encoder activation, $W \in \mathbb{R}^{F \times G}$ is the weight matrix, and $b_e \in \mathbb{R}^F$ is the encoder bias. $y = f_d(h) = a_d(W^T h)$ span the latent features back to the original space and y is the counterpart of x and a_d is the activation for the decoder. Since we use a deep autoencoder encoder and decoder functions are expanded for multilayer as $h_i = f_{e_i}(\ldots f_{e_2}(f_{e_1}(x))) = a_{e_i}(\ldots a_{e_2}(W_2 a_{e_1}(W_1 x + b_{e_1}) + b_{e_2}))$ $y_i = f_{d_1}(\ldots (f_{d_{i-1}}(f_{d_i}(h_i)))) = a_{d_1}(W_1^T \ldots (W_{i-2}^T a_{d_{i-1}}(W_{i-1}^T a_{d_i}(W_i^T h_i)h_{i-1})))$, respectively. The objective of an autoencoder is to minimize $\{W_i, b_{e_i}\}$:

$$\mathcal{J}_{AE} = \mathbb{E}_x\left[\ell\left(x, f_{d_i}\left(f_{e_i}(x)\right)\right)\right] \tag{6}$$

This captures the predominant patterns in the data and provides a noise invariant representation (manifold) of data which is very valuable considering the sensitivity of the infrared to noise. $\ell(.)$ denotes the loss function and here, we use binary cross-entropy (BCE), as presented below:

$$\mathcal{L}_{BCE} = -\frac{1}{F}\sum_{i=1}^{F} y_i log(p(y_i)) + (1 - y_i)\log(1 - p(y_i)) \tag{7}$$

where y is the label and $p(y)$ is the predicted probability of the segmented label for all F points. Having a Sigmoid function, $\frac{1}{1+e^{-y}}$ makes the function a binarized value, representing the existing class against background class.

Learning a dictionary fitted to a training set with the sparse latent code is formulated by the optimization below [89,90]:

$$min_{W_i, h_i} \sum_{j=1}^{F}\left(\| x_j - W_j^T h_j \|^2 + \lambda \| h_j \|_1\right) \tag{8}$$

This is a convex objective in every W_i and h_i when the other is fixed. ℓ_1 penalty term is the driving force in the above object forces for the sparse latent variable [88]. Here, the aforementioned objective is implemented for W_1 and h_1. Having sparse distributed representation (SDR) in this autoencoder not only follows the fundamental direction of deep learning but also creates robustness against noise [91]. Having our data compressed, we use a random forest to stratify the participants based on the sparse-latent deep features (Figure 3).

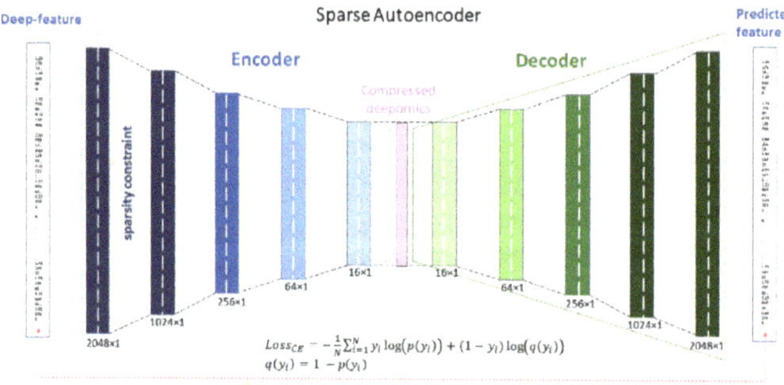

Figure 3. The proposed sparse deep autoencoder to reduce the dimensionality of the deep-thermomics is presented.

5. Results

The proposed method for thermal pattern detection was examined by thermal breast cancer screening datasets. The results of the low-rank approximation using sparse PCT were then compared to PCT thermal low-rank matrix approximation algorithm, as commonly used in infrared diagnostic systems.

5.1. Patient Population and Infrared Breast Cancer Study Data

We used 208 participants from Database for Mastology Research (DMR) infrared breast screening dataset [92], who were healthy (without symptoms) or sick (diagnosed by mammographic imaging as breast cancer cases or non-cancerous but with symptoms). The median age in our study sample was 60 years, and the participants comprised 77 (37%) Caucasian, 57 African (27.4%), 72 Pardo (34.6%), 1 Mulatto (0.5%), and 1 indigenous (0.5%) women. Among the participants, 52 had a history of diabetes in their families (25%), and 38 were undergoing hormone replacement (18.3%). All patients had infrared images obtained by the following acquisition protocol: images have a spatial resolution of 640 × 480 pixels and were captured by a FLIR thermal camera (model SC620) with a sensitivity of less than 0.04 °C range and capture standard of −40 °C to 500 °C [27,92]. Table 1 shows the clinical information and demography of the cohort. In this study, we considered symptomatic patients (who are not diagnosed with cancer but have similar signs) and sick (cancer) patients in one group, called the symptomatic group. The rationale behind this is due to having such analyses as the first line of screening, and once heterogeneity is detected by this system, further investigations need to be performed by a physician and another imaging modalities, i.e., mammography, to confirm the malignancy and specify the possible type of tumor.

Table 1. Clinical information and demographics of the breast cancer screening database using thermal imaging.

DMR—Database for Mastology Research		
Age	Median (±IQR)	60 (25,120)
Race	Caucasian	77 (37%)
	African	57 (27.4%)
	Pardo	72 (34.6%)
	Mulatto	1 (0.5%)
	Indigenous	1 (0.5%)
Diagnosis [1]	Healthy [2]	128 (61.5%)
	Symptomatic (with and without cancer)	80 (38.5%)
	Sick [3]	36 (17.3%)
Family history	Diabetes	52 (25%)
	Hypertensive	5 (2.4%)
	Leukemia	1 (0.5%)
	None	150 (72.1%)
Hormone therapy (HT)	Hormone replacement	38 (18.3%)
	None	170 (81.7%)

[1] This diagnosis performed with mammography as ground truth in this Dataset. [2] Healthy term is used as non-cancerous and non-symptomatic patients. [3] We use the term "sick", which includes different types of breast cancer patients diagnosed by mammographic imaging.

5.2. Results of Low-Rank Sparse PCT (Principal Component Analysis)

Three low-rank matrices were extracted from the 23 initially thermal sequences by using Sparse PCT. Some representative results of the low-rank approximation manually selected for our study cohort are shown in Figure 4. Low-rank approximation in the sequence of thermal images resulted in a heterogeneous breast area for 80 participants for breast cancer screening (sick and healthy with symptoms versus completely healthy without any symptom, Figure 4a–c). Thermal patterns showed more heterogeneous textures presenting the vasodilatory effect on the subdermal area of the breast. However, there was much less thermal heterogeneity found among the healthy participants (Table 1, Figure 4d–f). The targeted areas indicate significantly lesser heterogeneous patterns projected by the low-rank Sparse PCA in the ROI.

5.3. Deep-Thermomic Features

We extracted 2048 deep thermomic features from the targeted ROI in the thermal imaging (solely breasts area) using the latest layer of the ResNet-50 pre-trained model. The ResNet-50 pre-trained model contained five identical blocks having a convolutional layer, max-pooling, ReLu, and many repetitive identity connections between each layer. Convolutional blocks consisted of three convolution layers like identity block. The ResNet-50 model had 25,583,592 trainable parameters and 53,120 non-trainable parameters. The preferred input image to ResNet-50 was an RGB image with a dimension of $224 \times 224 \times 3$. This squared image slipped through the entire process and ResNet-50 model re-scaled the spatial dimensions of the input image from 224 to 230, 112, 56, 28, 14, 7 while the fourth dimension grew from 3 to 64, 128, 256, 512, and 2048. Here, we used this to leverage the low-rank matrix approximation for each participant. We extracted three bases using sparse low-rank matrix approximations from the original thermal stream and stacked them like three channels in the input image. The input image was cropped around the ROI to create a square matrix for each channel identically.

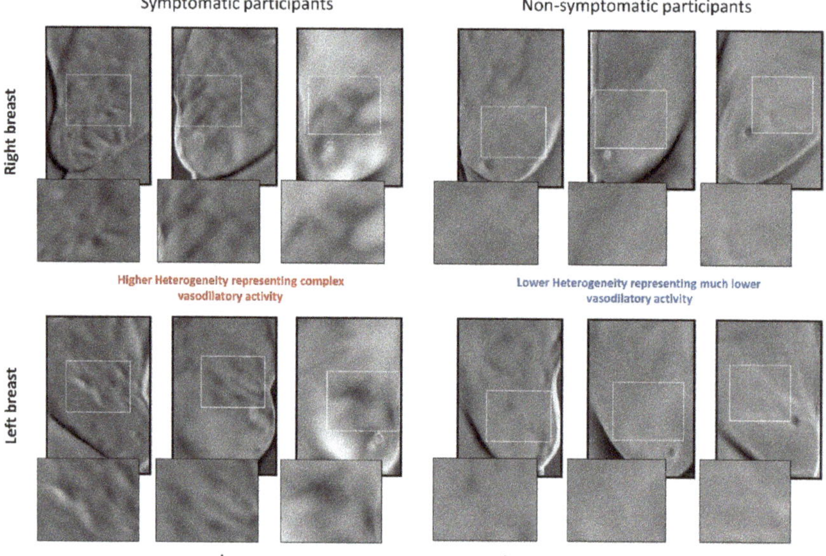

Figure 4. Low-rank approximation of thermal sequence determined using different Sparse PCT (principal component analysis) matrix factorization technique. Each column shows different case, columns (**a–c**) show symptomatic patients (diagnosed by mammography as cancer patients or healthy with symptoms), whereas columns (**d–f**) show the result of methods for healthy cases.

5.4. Result of the Sparse Autoencoder and Dimensionality Reduction

We used 2048 deep features extracted from the ResNet-50 pre-trained model as the input of the proposed autoencoder. The autoencoder consisted of seven layers (Figure 3) and reduced the dimensionality from 2048 to 1024, 256, 64, and 16. The model was trained and validated with 4000, and 2000 vectors obtained by the ResNet-50 model from infrared images in the breast screening dataset. The model had 4,744,784 trainable parameters and was trained by *Adam* optimizer with the learning ℓ_1 with a regularization value of 10^{-5} and for 500 epochs. The batch size was 128 for the model network. Figure 5 shows the loss of the model during the training.

From 2048 initially extracted deep thermomic features, the extracted features from the battle-neck layer of autoencoder resulted in 16 deep-thermomic features (Figure 5). Subsequently, the level of the heterogeneity for each participant measured through compressed descriptors was obtained by 128 times compression on the deep-thermomics.

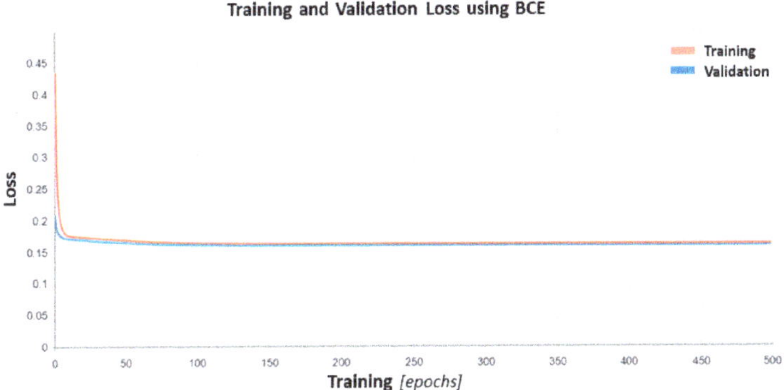

Figure 5. The binary cross entropy loss is presented for training and validation of the proposed sparse autoencoder for 300 epochs.

5.5. Result of Random Forest Classification of Symptomatic and Non-Symptomatic Participants

We stratified the participants based on the 16 sparse latent deep thermomic descriptors and compared them with the ground truth data based on mammography information. To examine the hypothesis that the thermal heterogeneity extracted by the deep learning model can be used as a biomarker to stratify among participants, a random forest classifier was fitted for multivariate covariates with leave-one-out cross-validation. The best multivariate model resulted accuracy of 75.24% (72.33–77.67%) for Sparse PCT, which was challenged by other matrix approximation technique PCT (73.27% (71.84–76.21%)). A multivariate model contains clinical information and demographics (age, and family history) gave an accuracy of 71.36% (69.42–73.3%). A full multivariate model having all clinical and demographic information with the extracted features resulted in 78.16% (73.3–81.07%) for Sparse PCT and 73.79% (72.33–76.7%) for PCT (see Table 2). The receiver operating characteristic (ROC) curve of comparative analyses of baseline models is shown in Figure 6. The entire computational experiments were conducted by Python programming language [93] (for training and testing the model).

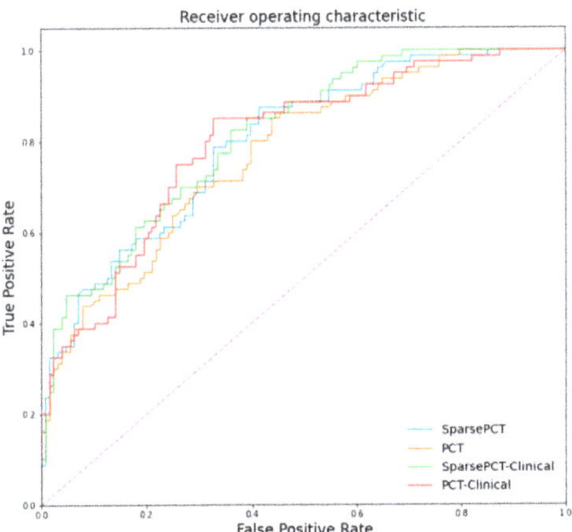

Figure 6. Receiver operating characteristic (ROC) curve for different multivariate model using deep thermomic features and clinical and demographic information is presented for classifying between symptomatic and non-symptomatic participants (for baseline model).

Table 2. The results of random forest classification for the cross-validated model.

Methods	Cross-Validated Accuracy
Sparse PCT	75.24 (72.33–77.67)%
PCT	73.27 (71.84–76.21)%
Clinical information *	71.36 (69.42–73.3)%
Sparse PCT with clinical information *	78.16 (73.3–81.07)%
PCT with clinical information *	73.79 (72.33–76.7)%

* Clinical and demographic covariates: age, and family history.

6. Discussion

In this study, we proposed a system to reduce the dimensionality of deep-thermomic features to extract thermal patterns for infrared diagnostic systems for thermography imaging. This study was designed based on the general trend of dimensionality reduction and to alleviate the possibility of over-fitting but used sparse multiple low-rank matrix approximations. This study showed a possibility to identify potential patients with breast cancer along with other clinical throughputs (such as CBE) using non-invasive, faster, and more cost-efficient thermography imaging.

The application of deep sparse autoencoder not only reduced the initial high-dimensional deep-thermomics but also added sparsity to the initial sparse representative of the thermal stream, which theoretically increases the robustness of the system against noise. It also showed significant improvements in stratifying symptomatic patients from healthy participants (Figure 6, and Table 2). Moreover, using sparse PCT showed higher accuracy than other approaches in finding heterogeneous thermal patterns, which might be due to the nature of sparsity in the calculation of the low-rank representative of the basis matrices, which were preserved by ResNet-50 level-level features and recursive training of the autoencoder network. This indicates the penalty terms in Sparse PCT creates constraints that worked in favor of detecting symptomatic cases while eliminating noise.

The application of deep thermomics considerably increased the dimensionality of the input thermal imaging and intensify the possibility of overfitting the random forest model, called the *curse of dimensionality*. The proposed sparse autoencoder reduced the dimensionality by removing the redundancy among the features by spanning thermomics to lower-dimensional space, while increasing the robustness of feature selection due to rigorous training of the model (this method is also used in other applications in medicine such as segmentation [94]). Since the infrared images used in this study had intensity information similar to natural images, using this pre-trained model is seemingly appropriate despite the medical nature of the analysis.

Thermal and infrared imagery has been used to determine breast abnormality, as the first medical application of thermography [92]. There are many discussions about more suitable positions for such imaging acquisitions [35] and the reliability of this modality [57,58] that have been reported. However, the association of sparse autoencoder on the abundant deep thermomics finding thermal heterogeneity with a breast abnormality has not been discussed in literature which increases the novelty of this contribution to the field. One of the reasons that the proposed system performs well finding thermal heterogeneity might be because of the association of low-level deep features representing basis set and their sensitivity to slight intensity variation of thermal images. This can be justified by vasodilatory activity in the ROI for symptomatic patients. Despite some argument against using a thermal infrared imaging system as a solo-imaging modality for detecting breast abnormality, this technology has still been used as one of the important diagnostic tools with CBE and other imaging modalities (as discussed in Section 2).

One limitation to applying the presented models is related to data, and even with a considerable number of cases there is a need to increase the cohort to confirm the accuracy of the system. Having a larger cohort of patients increases the statistical power of such analysis by increasing the possibility of independently validating the system (substitute cross-validation). The other limitation may be using limited deep thermomic features. Having more deep-thermomics helps to assess the strength of the sparse autoencoder approach to select better compressing features that lead to capturing better thermal characteristics. There is an inherent limitation of using infrared thermography for detecting cancerous tissues once they are deeper into the tissues, as despite tracking the vasodilatory effects in the skin subsurface, deeper lesions might not be easy to detect. This might require further investigation using multimodal imaging analyses. There are also discussions on using a single thermal frame or multiple frames for selecting variability of thermal patterns across the acquisition time using different infrared-based approaches [5,35,95–97], which can be investigated further. Applying multiple thermal frames provided a chance of capturing thermal heterogeneity in the ROI for the duration of the acquisition, which might not be recorded by a single frame input system.

The presented technique offers some advantages. First, applying low-rank matrix approximation to extract thermal *avatars* during imaging acquisition provides a significant projection of thermal heterogeneity leading to better diagnosis of abnormal patients. Second, a sparse autoencoder eliminates the manual selection or human-engineering feature selection for reducing the dimensionality of the deep thermomics. Third, the proposed method considerably alleviates the effect of motion artifacts and imaging acquisition noise, which can be substantial improvements in infrared thermography applications. To the best of our knowledge, this is the first study which performs such analyses.

7. Conclusions

This study addressed one of the biggest challenges in high-dimensional deep feature selection, which selected the best representative deep thermomics from high-dimensional features extracted from a pre-trained deep neural networks model. The method performed multilayer dimensionality reduction using Sparse PCT to select the low-rank approximation of the thermal sequence. They extracted high-dimensional features from the ResNet-50 pre-trained model. Then, it used a trained sparse autoencoder to hierarchically reduce the size of the feature to 16 thermomic descriptors. We tested our method for 208 thermal breast cancer screening cases. We compared the appropriateness of these approaches with similar state-of-the-art thermographic methods, i.e., PCT. The results indicated the significant performance of the full multivariate model using Sparse PCT in preserving thermal heterogeneity to discriminate between symptomatic and healthy participants (accuracy of 78.16% (73.3–81.07%)).

Future works should involve more thermomics extracted from the different low-rank approximations to increase the potential of assessing the entire thermal characteristics of cancerous parenchymal tissues. Moreover, an expansion of the validation set to a larger infrared imaging cohort can further confirm the strength and limitations of this approach.

Author Contributions: Conceptualization, B.Y. and H.A.; methodology, software, validation, formal analysis, investigation, resources, B.Y.; writing—original draft preparation, B.Y.; writing—review and editing, B.Y., H.A., X.P.V.M.; visualization, B.Y.; supervision, X.P.V.M. All authors have read and agreed to the published version of the manuscript.

Funding: This research received no external funding.

Acknowledgments: This research is conducted under a tier-one—Canadian Research Chair in Multipolar Infrared Vision (MIVIM) at the Department of Electrical and Computer Engineering at Laval University.

Conflicts of Interest: The authors declare no conflict of interest.

References

1. Siegel, R.L.; Miller, K.D.; Jemal, A. Cancer statistics, 2019. *CA Cancer J. Clin.* **2019**, *69*, 7–34. [CrossRef] [PubMed]
2. Crystal, P.; Strano, S.D.; Shcharynski, S.; Koretz, M.J. Using Sonography to Screen Women with Mammographically Dense Breasts. *Am. J. Roentgenol.* **2003**, *181*, 177–182. [CrossRef]
3. Kerlikowske, K.; Grady, D.; Barclay, J.; Sickles, E.A.; Ernster, V. Effect of age, breast density, and family history on the sensitivity of first screening mammography. *JAMA* **1996**, *276*, 33–38. [CrossRef] [PubMed]

4. Hong, S.; Song, S.Y.; Park, B.; Suh, M.; Choi, K.S.; Jung, S.E.; Kim, M.J.; Lee, E.H.; Lee, C.W.; Jun, J.K. Effect of Digital Mammography for Breast Cancer Screening: A Comparative Study of More than 8 Million Korean Women. *Radiology* **2020**, *294*, 247–255. [CrossRef]

5. Kennedy, D.A.; Lee, T.; Seely, D. A comparative review of thermography as a breast cancer screening technique. *Integr. Cancer Ther.* **2009**, *8*, 9–16. [CrossRef]

6. Fletcher, S.W.; Elmore, J.G. Mammographic Screening for Breast Cancer. *N. Engl. J. Med.* **2003**, *348*, 1672–1680. [CrossRef] [PubMed]

7. Osako, T.; Iwase, T.; Takahashi, K.; Lijima, K.; Miyagi, Y.; Nishimura, S.; Tada, K.; Makita, M.; Akiyama, F.; Sakamoto, G.; et al. Diagnostic mammography and ultrasonography for palpable and nonpalpable breast cancer in women aged 30 to 39 years. *Breast Cancer* **2007**, *14*, 255–259. [CrossRef] [PubMed]

8. Duffy, S.W.; Vulkan, D.; Cuckle, H.; Parmar, D.; Sheikh, S.; Smith, R.A.; Evans, A.; Blyuss, O.; Johns, L.; Ellis, I.O.; et al. Effect of mammographic screening from age 40 years on breast cancer mortality (UK Age trial): Final results of a randomised, controlled trial. *Lancet Oncol.* **2020**, *21*, 1165–1172. [CrossRef]

9. Nguyen, Q.D.; Nguyen, N.T.; Dixon III, L.; Monetto, F.E.P.; Robinson, A.S. Spontaneously Disappearing Calcifications in the Breast: A Rare Instance Where a Decrease in Size on Mammogram Is Not Good. *Cureus* **2020**, *12*, e8753. [CrossRef]

10. De González, A.B.; Reeves, G. Mammographic screening before age 50 years in the UK: Comparison of the radiation risks with the mortality benefits. *Br. J. Cancer* **2005**, *93*, 590–596. [CrossRef] [PubMed]

11. Hoekstra, P. Quantitive digital thermology: 21st century imaging systems. In Proceedings of the OAND Conference, Hamilton, ON, Canada, 18–21 June 2001.

12. Law, J.; Faulkner, K.; Young, K.C. Risk factors for induction of breast cancer by X-rays and their implications for breast screening. *Br. J. Radiol.* **2007**, *80*, 261–266. [CrossRef] [PubMed]

13. Hendrick, R.E. Radiation Doses and Risks in Breast Screening. *J. Breast Imaging* **2020**, *2*, 188–200. [CrossRef]

14. Chowdhury, A.S.; Tamanna, S. Radiation-induced side effects in breast cancer patients and factors affecting them. *Asian J. Med Biol. Res.* **2020**, *6*, 138–148. [CrossRef]

15. Nelson, H.D.; Pappas, M.; Cantor, A.; Griffin, J.; Daeges, M.; Humphrey, L. Harms of breast cancer screening: Systematic review to update the 2009 US Preventive Services Task Force recommendation. *Ann. Intern. Med.* **2016**, *164*, 256–267. [CrossRef]

16. Bandyopadhyay, U.; Das, D.; Banerjee, R.K. Reactive oxygen species: Oxidative damage and pathogenesis. *Curr. Sci.* **1999**, *77*, 658–666.

17. Yang, Y.; Bazhin, A.V.; Werner, J.; Karakhanova, S. Reactive Oxygen Species in the Immune System. *Int. Rev. Immunol.* **2013**, *32*, 249–270. [CrossRef]

18. Friedenson, B. Is mammography indicated for women with defective BRCA genes? Implications of recent scientific advances for the diagnosis, treatment, and prevention of hereditary breast cancer. *MedGenMed* **2000**, *2*, E9. [PubMed]

19. Narod, S.A.; Lubinski, J.; Ghadirian, P.; Lynch, H.T.; Moller, P.; Foulkes, W.; Rosen, B.; Kim-Sing, C.; Isaacs, C.; Domcheck, S.; et al. Screening mammography and risk of breast cancer in BRCA1 and BRCA2 mutation carriers: A case-control study. *Lancet Oncol.* **2006**, *7*, 402–406. [CrossRef]

20. Khalkhali, I.; Vargas, H. Practical use of ultrasound at a dedicated breast center. *Breast J.* **2005**, *11*, 165–166. [CrossRef]

21. Burkett, B.J.; Hanemann, C.W. A Review of Supplemental Screening Ultrasound for Breast Cancer. *Acad. Radiol.* **2016**, *23*, 1604–1609. [CrossRef]

22. Berg, W.A.; Zhang, Z.; Lehrer, D.; Jong, R.A.; Pisano, E.D.; Barr, R.G.; Böhm-Vélez, M.; Mahoney, M.C.; Evans, W.P.; Larsen, L.H.; et al. Detection of Breast Cancer With Addition of Annual Screening Ultrasound or a Single Screening MRI to Mammography in Women With Elevated Breast Cancer Risk. *JAMA* **2012**, *307*, 1394–1404. [CrossRef]

23. Morris, E.A. Screening for breast cancer with MRI. In Seminars in Ultrasound, CT and MRI. *WB Saunders* **2003**, *24*, 45–54.

24. Ibrahim, A.; Mohammed, S.; Ali, H.A. Breast Cancer Detection and Classification Using Thermography: A Review. In *Proceedings of the International Conference on Advanced Machine Learning Technologies and Applications, Cairo, Egypt, 22–24 February 2018*; Springer: Cham, Switzerland, 2018; pp. 496–505.

25. Ahern, C.H.; Shih, Y.-C.T.; Dong, W.; Parmigiani, G.; Shen, Y. Cost-effectiveness of alternative strategies for integrating MRI into breast cancer screening for women at high risk. *Br. J. Cancer* **2014**, *111*, 1542–1551. [CrossRef]

26. Barton, M.B.; Harris, R.; Fletcher, S.W. Does This Patient Have Breast Cancer? *JAMA* **1999**, *282*, 1270–1280. [CrossRef]

27. Oestreicher, N.; Lehman, C.D.; Seger, D.J.; Buist, D.S.M.; White, E. The Incremental Contribution of Clinical Breast Examination to Invasive Cancer Detection in a Mammography Screening Program. *Am. J. Roentgenol.* **2005**, *184*, 428–432. [CrossRef]

28. Freed, D.; Golden, J.; Chu, M.; Carrillo, O.; Chin, Y.; Adams, M.; Abramov, V. Medical Devices and Systems. U.S. Patent Application No.11/388,247, 23 March 2006.

29. Kirimtat, A.; Krejcar, O.; Selamat, A.; Herrera-Viedma, E. FLIR vs SEEK thermal cameras in biomedicine: Comparative diagnosis through infrared thermography. *BMC Bioinform.* **2020**, *21*, 1–10. [CrossRef] [PubMed]

30. Moore, G. Breast cancer: Early detection needed. *Bus. Heal.* **2001**, *19*, 39.

31. Feig, S.A.; Shaber, G.S.; Schwartz, G.F.; Patchefsky, A.; Libshitz, H.I.; Edeiken, J.; Nerlinger, R.; Curley, R.F.; Wallace, J.D. Thermography, Mammography, and Clinical Examination in Breast Cancer Screening. *Radiology* **1977**, *122*, 123–127. [CrossRef]

32. Keyserlingk, J.; Ahlgren, P.; Yu, E.; Belliveau, N.; Yassa, M. Functional infrared imaging of the breast. *IEEE Eng. Med. Boil. Mag.* **2000**, *19*, 30–41. [CrossRef] [PubMed]

33. Anbar, M. Clinical thermal imaging today. *IEEE Eng. Med. Boil. Mag.* **1998**, *17*, 25–33. [CrossRef] [PubMed]

34. Gautherie, M. Thermopathology of breast cancer: Measurement and analysis of in vivo temperature and blood flow. *Ann. N. Y. Acad. Sci.* **1980**, *335*, 383–415. [CrossRef]

35. Recinella, A.N.; Gonzalez-Hernandez, J.-L.; Kandlikar, S.G.; Dabydeen, D.; Medeiros, L.; Phatak, P. Clinical Infrared Imaging in the Prone Position for Breast Cancer Screening—Initial Screening and Digital Model Validation. *J. Eng. Sci. Med. Diagn. Ther.* **2020**, *3*. [CrossRef]

36. Mc Donald, D. Mechanism of Tumour Leakiness Proceeding Angiogenesis and Cancer. From Basic Mechanisms to Therapeutic Applications. In Proceedings of the American Association of Cancer Research Conference (AACR), San Francisco, CA, USA, 1–5 April 2000; pp. 11–15.

37. Gamagami, P. Indirect Signs of Breast Cancer: Angiogenesis Study. In *Atlas of Mammography*; Blackwell Science: Cambridge, MA, USA, 1996; pp. 321–326.

38. Yoshida, S.; Nakagawa, S.; Yahara, T.; Koga, T.; Deguchi, H.; Shirouzu, K. Relationship Between Microvessel Density and Thermographic Hot Areas in Breast Cancer. *Surg. Today* **2003**, *33*, 243–248. [CrossRef]

39. Mitra, S.; Bal, A.; Kashyap, D.; Kumar, S.; Shrivastav, S.; Das, A.; Singh, G. Tumour angiogenesis and c-Met pathway activation—Implications in breast cancer. *APMIS* **2020**, *128*, 316–325. [CrossRef]

40. Tsutsui, S.; Yasuda, K.; Suzuki, K.; Tahara, K.; Higashi, H.; Era, S. Macrophage infiltration and its prognostic implications in breast cancer: The relationship with VEGF expression and microvessel density. *Oncol. Rep.* **2005**, *14*, 425–431. [CrossRef]

41. Buckley, D.L.; Drew, P.J.; Mussurakis, S.; Monson, J.R.; Horsman, A. Microvessel density in invasive breast cancer assessed by dynamic gd-dtpa enhanced MRI. *J. Magn. Reson. Imaging* **1997**, *7*, 461–464. [CrossRef]

42. Anbar, M.; Brown, C.; Milescu, L.; Babalola, J.; Genther, L. The potential of dynamic area telethermometry in assessing breast cancer. *IEEE Eng. Med. Boil. Mag.* **2000**, *19*, 58–62. [CrossRef] [PubMed]

43. Thomsen, L.L.; Miles, D.W.; Happerfield, L.; Bobrow, L.G.; Knowles, R.G.; Moncada, S. Nitric oxide synthase activity in human breast cancer. *Br. J. Cancer* **1995**, *72*, 41–44. [CrossRef]

44. Arnedos, M.; Vicier, C.; Loi, S.; Lefebvre, C.; Michiels, S.; Bonnefoi, H.; Andre, F. Precision medicine for metastatic breast cancer—Limitations and solutions. *Nat. Rev. Clin. Oncol.* **2015**, *12*, 693. [CrossRef]

45. Di Meo, S.; Reed, T.T.; Venditti, P.; Victor, V.M. Role of ROS and RNS Sources in Physiological and Pathological Conditions. *Oxidative Med. Cell. Longev.* **2016**, *2016*, 1–44. [CrossRef]

46. Moldogazieva, N.T.; Lutsenko, S.V.; Terentiev, A.A. Reactive Oxygen and Nitrogen Species–Induced Protein Modifications: Implication in Carcinogenesis and Anticancer Therapy. *Cancer Res.* **2018**, *78*, 6040–6047. [CrossRef] [PubMed]

47. Smith, K.A.; Waypa, G.B.; Schumacker, P.T. Redox signaling during hypoxia in mammalian cells. *Redox Biol.* **2017**, *13*, 228–234. [CrossRef]

48. Xu, W.; Liu, L.Z.; Loizidou, M.; Ahmed, M.J.U.; Charles, I.G. The role of nitric oxide in cancer. *Cell Res.* **2002**, *12*, 311–320. [CrossRef]

49. Choudhari, S.K.; Chaudhary, M.; Bagde, S.; Gadbail, A.R.; Joshi, V. Nitric oxide and cancer: A review. *World J. Surg. Oncol.* **2013**, *11*, 1–11. [CrossRef]

50. Martínez, M.C.; Andriantsitohaina, R. Reactive Nitrogen Species: Molecular Mechanisms and Potential Significance in Health and Disease. *Antioxidants Redox Signal.* **2009**, *11*, 669–702. [CrossRef]

51. Xing, F.; Liu, Y.; Sharma, S.; Wu, K.; Chan, M.D.; Lo, H.-W.; Carpenter, R.L.; Metheny-Barlow, L.J.; Zhou, X.; Qasem, S.A.; et al. Activation of the c-Met Pathway Mobilizes an Inflammatory Network in the Brain Microenvironment to Promote Brain Metastasis of Breast Cancer. *Cancer Res.* **2016**, *76*, 4970–4980. [CrossRef]

52. Changchun, K.; Pengchao, H.; Ke, S.; Ying, W.; Wei, L. Interleukin-17 augments tumor necrosis factor α-mediated increase of hypoxia-inducible factor-1α and inhibits vasodilator-stimulated phosphoprotein expression to reduce the adhesion of breast cancer cells. *Oncol. Lett.* **2017**, *13*, 3253–3260. [CrossRef]

53. Maher, E.A.; Mietz, J.; Arteaga, C.L.; DePinho, R.A.; Mohla, S. Brain metastasis: Opportunities in basic and translational research. *Cancer Res.* **2009**, *69*, 6015–6020. [CrossRef]

54. Ganong, W.F. *Review of Medical Physiology*, 2nd ed.; Lange Medical Books/McGraw-Hill: San Francisco, CA, USA, 2005.

55. Spychala, J.; Lazarowski, E.; Ostapkowicz, A.; Ayscue, L.H.; Jin, A.; Mitchell, B.S. Role of estrogen receptor in the regulation of ecto-5′-nucleotidase and adenosine in breast cancer. *Clin. Cancer Res.* **2004**, *10*, 708–717. [CrossRef]

56. Lozano, A.; Hayes, J.C.; Compton, L.M.; Azarnoosh, J.; Hassanipour, F. Determining the thermal characteristics of breast cancer based on high-resolution infrared imaging, 3D breast scans, and magnetic resonance imaging. *Sci. Rep.* **2020**, *10*, 1–14. [CrossRef] [PubMed]

57. FDA Safety Communication, FDA Warns Thermography Should Not Be Used in Place of Mammography to Detect, Diagnose, or Screen for Breast Cancer. 2019. Available online: www.fda.gov (accessed on 25 February 2019).

58. FDA Consumer Update, Breast Cancer Screening: Thermogram No Substitute for Mammogram, U.S. Food and Drug Administration. 2017. Available online: www.fda.gov (accessed on 30 October 2017).

59. Pennes, H.H. Analysis of Tissue and Arterial Blood Temperatures in the Resting Human Forearm. *J. Appl. Physiol.* **1948**, *1*, 93–122. [CrossRef]

60. Rajic, N. Principal component thermography for flaw contrast enhancement and flaw depth characterisation in composite structures. *Compos. Struct.* **2002**, *58*, 521–528. [CrossRef]

61. Yousefi, B.; Sfarra, S.; Ibarra-Castanedo, C.; Maldague, X.P. Comparative analysis on thermal non-destructive testing imagery applying Candid Covariance-Free Incremental Principal Component Thermography (CCIPCT). *Infrared Phys. Technol.* **2017**, *85*, 163–169. [CrossRef]

62. Yousefi, B.; Sfarra, S.; Sarasini, F.; Castanedo, C.I.; Maldague, X.P. Low-rank sparse principal component thermography (sparse-PCT): Comparative assessment on detection of subsurface defects. *Infrared Phys. Technol.* **2019**, *98*, 278–284. [CrossRef]

63. Wu, J.-Y.; Sfarra, S.; Yao, Y. Sparse Principal Component Thermography for Subsurface Defect Detection in Composite Products. *IEEE Trans. Ind. Inform.* **2018**, *14*, 5594–5600. [CrossRef]

64. Usamentiaga, R.; Mokhtari, Y.; Ibarra-Castanedo, C.; Klein, M.; Genest, M.; Maldague, X. Automated Dynamic Inspection Using Active Infrared Thermography. *IEEE Trans. Ind. Inform.* **2018**, *14*, 5648–5657. [CrossRef]

65. Marinetti, S.; Finesso, L.; Marsilio, E. Matrix factorization methods: Application to thermal NDT/E. *NDT E Int.* **2006**, *39*, 611–616. [CrossRef]

66. Cramer, K.E.; Winfree, W. Fixed eigenvector analysis of thermographic NDE data. *SPIE* **2011**, 225–235. [CrossRef]

67. Yousefi, B.; Sharifipour, H.M.; Eskandari, M.; Castanedo, C.I.; Laurendeau, D.; Watts, R.; Klein, M.; Maldague, X. Incremental Low Rank Noise Reduction for Robust Infrared Tracking of Body Temperature during Medical Imaging. *Electronics* **2019**, *8*, 1301. [CrossRef]

68. Ahmed, J.; Gao, B.; Woo, W.L. Wavelet-Integrated Alternating Sparse Dictionary Matrix Decomposition in Thermal Imaging CFRP Defect Detection. *IEEE Trans. Ind. Inform.* **2018**, *15*, 4033–4043. [CrossRef]

69. Zou, H.; Hastie, T.; Tibshirani, R. Sparse principal component analysis. *J. Comput. Graph. Stat.* **2006**, *15*, 265–286. [CrossRef]

70. FeCun, Y. LeNet-5, Convolutional Neural Networks. 2015. Available online: http://yann.lecun.com/exdb/lenet (accessed on 31 October 2020).

71. Aghdam, H.H.; Heravi, E.J. *Guide to Convolutional Neural Networks: A Practical Application to Traffic-Sign Detection and Classification*; Springer: Berlin/Heidelberg, Germany, 2017.

72. Krizhevsky, A.; Sutskever, I.; Hinton, G.E. ImageNet classification with deep convolutional neural networks. In Proceedings of the Advances in Neural Information Processing Systems (NIPS), Lake Tahoe, NV, USA, 3–6 December 2012; pp. 1097–1105.

73. He, K.; Zhang, X.; Ren, S.; Sun, J. Deep Residual Learning for Image Recognition. In Proceedings of the 2016 IEEE Conference on Computer Vision and Pattern Recognition (CVPR), Las Vegas, NV, USA, 26 June–1 July 2016; pp. 770–778.

74. Simonyan, K.; Zisserman, A. Very deep convolutional networks for large-scale image recognition. *arXiv* **2014**, arXiv:1409.1556.

75. Szegedy, C.; Liu, W.; Jia, Y.; Sermanet, P.; Reed, S.; Anguelov, D.; Erhan, D.; Vanhoucke, V.; Rabinovich, A. Going deeper with convolutions. In Proceedings of the 2015 IEEE Conference on Computer Vision and Pattern Recognition (CVPR), Boston, MA, USA, 7–12 June 2015; pp. 1–9.

76. He, K.; Zhang, X.; Ren, S.; Sun, J. Identity Mappings in Deep Residual Networks. *arXiv* **2016**, arXiv:1603.05027v3.

77. Liang, L.; Liu, M.; Sun, W. A deep learning approach to estimate chemically-treated collagenous tissue nonlinear anisotropic stress-strain responses from microscopy images. *Acta Biomater.* **2017**, *63*, 227–235. [CrossRef]

78. Akbari, H.; Rathore, S.; Bakas, S.; Nasrallah, M.P.; Shukla, G.; Mamourian, E.; Rozycki, M.; Bagley, S.J.; Rudie, J.D.; Flanders, A.E.; et al. Histopathology-validated machine learning radiographic biomarker for noninvasive discrimination between true progression and pseudo-progression in glioblastoma. *Cancer* **2020**, *126*, 2625–2636. [CrossRef]

79. Yousefi, B.; Kalhor, D.; Usamentiaga, R.; Lei, L.; Ibarra-Castanedo, C.; Maldague, X. Application of Deep Learning in Infrared Non-Destructive Testing. *Quant. InfraRed Thermogr.* **2018**. [CrossRef]

80. Chaves, E.; Gonçalves, C.B.; Albertini, M.K.; Lee, S.; Jeon, G.; Fernandes, H.C. Evaluation of transfer learning of pre-trained CNNs applied to breast cancer detection on infrared images. *Appl. Opt.* **2020**, *59*, E23–E28. [CrossRef]

81. Acharya, S.; Alsadoon, A.; Prasad, P.W.C.; Abdullah, S.; Deva, A. Deep convolutional network for breast cancer classification: Enhanced loss function (ELF). *J. Supercomput.* **2020**, *76*, 1–18. [CrossRef]

82. Fernández-Ovies, F.J.; Alférez-Baquero, E.S.; De Andrés-Galiana, E.J.; Cernea, A.; Fernández-Muñiz, Z.; Fernández-Martínez, J.L. Detection of Breast Cancer Using Infrared Thermography and Deep Neural Networks. In *Proceedings of the International Work-Conference on Bioinformatics and Biomedical Engineering, Granada, Spain, 8–10 May 2019*; Springer: Cham, Switzerland, 2019; pp. 514–523.

83. Zuluaga-Gomez, J.; Al Masry, Z.; Benaggoune, K.; Meraghni, S.; Zerhouni, N. A CNN-based methodology for breast cancer diagnosis using thermal images. *Comput. Methods Biomech. Biomed. Eng. Imaging Vis.* **2020**, 1–15. [CrossRef]

84. Ekici, S.; Jawzal, H. Breast cancer diagnosis using thermography and convolutional neural networks. *Med. Hypotheses* **2020**, *137*, 109542. [CrossRef]

85. Mambou, S.J.; Maresova, P.; Krejcar, O.; Selamat, A.; Kuca, K. Breast Cancer Detection Using Infrared Thermal Imaging and a Deep Learning Model. *Sensors* **2018**, *18*, 2799. [CrossRef]

86. He, K.; Zhang, X.; Ren, S.; Sun, J. Deep Residual Learning for Image Recognition. *arXiv* **2015**, arXiv:1512.03385.

87. Gillies, R.J.; Kinahan, P.E.; Hricak, H. Radiomics: Images Are More than Pictures, They Are Data. *Radiology* **2016**, *278*, 563–577. [CrossRef]

88. Chollet, F. Building Autoencoders in Keras. 2016. Available online: https://blog.keras.io/building-autoencoders-in-keras.html (accessed on 14 March 2017).

89. Olshausen, B.A.; Field, D.J. Sparse coding with an overcomplete basis set: A strategy employed by V1? *Vis. Res.* **1997**, *37*, 3311–3325. [CrossRef]

90. Arpit, D.; Zhou, Y.; Ngo, H.; Govindaraju, V. Why regularized auto-encoders learn sparse representation? In Proceedings of the International Conference on Machine Learning, New York, NY, USA, 19–24 June 2016; pp. 136–144.

91. Hinton, G.E. Distributed Representations. 1984. Available online: https://web.stanford.edu/~jlmcc/papers/PDP/ (accessed on 18 July 2005).

92. Silva, L.F.; Saade, D.C.M.; Sequeiros, G.O.; Silva, A.C.; Paiva, A.C.; Bravo, R.S.; Conci, A. A New Database for Breast Research with Infrared Image. *J. Med Imaging Heal. Inform.* **2014**, *4*, 92–100. [CrossRef]
93. Python 3 Google Compute Engine Backend, T4, and P100 GPU and 27.4 Gb RAM, Google 2020.
94. Mortazi, A.; Khosravan, N.; Torigian, E.A.; Kurugol, S.; Bagci, U. Weakly Supervised Segmentation by a Deep Geodesic Prior. In *Lecture Notes in Computer Science*; Springer: Cham, Switzerland, 2019; pp. 238–246.
95. Yousefi, B.; Castanedo, C.I.; Maldague, X.P. Measuring heterogeneous thermal patterns in infrared-based diagnostic systems using sparse low-rank matrix approximation: Comparative study. *IEEE Trans. Instrum. Meas.* **2020**. [CrossRef]
96. Yousefi, B.; Castanedo, C.I.; Maldague, X.P.V. Low-rank Convex/Sparse Thermal Matrix Approximation for Infrared-based Diagnostic System. *arXiv* **2020**, arXiv:2010.06784.
97. Kirimtat, A.; Krejcar, O. A review of infrared thermography for the investigation of building envelopes: Advances and prospects. *Energy Build.* **2018**, *176*, 390–406. [CrossRef]

Publisher's Note: MDPI stays neutral with regard to jurisdictional claims in published maps and institutional affiliations.

Article

Towards Multi-Analyte Detection with Field-Effect Capacitors Modified with *Tobacco Mosaic Virus* Bioparticles as Enzyme Nanocarriers

Melanie Welden [1,2], Arshak Poghossian [3], Farnoosh Vahidpour [1], Tim Wendlandt [4], Michael Keusgen [2], Christina Wege [4] and Michael J. Schöning [1,5,*]

1   Institute of Nano- and Biotechnologies, Aachen University of Applied Sciences, 52428 Jülich, Germany; m.welden@fh-aachen.de (M.W.); vahidpour@fh-aachen.de (F.V.)
2   Institute of Pharmaceutical Chemistry, Philipps University Marburg, 35032 Marburg, Germany; michael.keusgen@staff.uni-marburg.de
3   MicroNanoBio, 40479 Düsseldorf, Germany; a.poghossian@gmx.de
4   Institute of Biomaterials and Biomolecular Systems, University of Stuttgart, 70569 Stuttgart, Germany; tim.wendlandt@bio.uni-stuttgart.de (T.W.); christina.wege@bio.uni-stuttgart.de (C.W.)
5   Institute of Biological Information Processing (IBI-3), Forschungszentrum Jülich GmbH, 52425 Jülich, Germany
*   Correspondence: schoening@fh-aachen.de

Abstract: Utilizing an appropriate enzyme immobilization strategy is crucial for designing enzyme-based biosensors. Plant virus-like particles represent ideal nanoscaffolds for an extremely dense and precise immobilization of enzymes, due to their regular shape, high surface-to-volume ratio and high density of surface binding sites. In the present work, *tobacco mosaic virus* (TMV) particles were applied for the co-immobilization of penicillinase and urease onto the gate surface of a field-effect electrolyte-insulator-semiconductor capacitor (EISCAP) with a p-Si-SiO$_2$-Ta$_2$O$_5$ layer structure for the sequential detection of penicillin and urea. The TMV-assisted bi-enzyme EISCAP biosensor exhibited a high urea and penicillin sensitivity of 54 and 85 mV/dec, respectively, in the concentration range of 0.1–3 mM. For comparison, the characteristics of single-enzyme EISCAP biosensors modified with TMV particles immobilized with either penicillinase or urease were also investigated. The surface morphology of the TMV-modified Ta$_2$O$_5$-gate was analyzed by scanning electron microscopy. Additionally, the bi-enzyme EISCAP was applied to mimic an **XOR (Exclusive OR)** enzyme logic gate.

Keywords: *tobacco mosaic virus* (TMV); capacitive field-effect sensor; bi-enzyme biosensor; enzyme-logic gate; urease; penicillinase

Citation: Welden, M.; Poghossian, A.; Vahidpour, F.; Wendlandt, T.; Keusgen, M.; Wege, C.; Schöning, M.J. Towards Multi-Analyte Detection with Field-Effect Capacitors Modified with *Tobacco Mosaic Virus* Bioparticles as Enzyme Nanocarriers. *Biosensors* 2022, 12, 43. https://doi.org/10.3390/bios12010043

Received: 13 December 2021
Accepted: 13 January 2022
Published: 14 January 2022

Publisher's Note: MDPI stays neutral with regard to jurisdictional claims in published maps and institutional affiliations.

1. Introduction

The field effect in an electrolyte-insulator-semiconductor (EIS) system offers a universal transducer principle for designing many kinds of chemical sensors and biosensors (see e.g., recent reviews [1–9]). EIS capacitors (EISCAP) are considered as the simplest type of such field-effect sensors [10]. They have been implemented for detecting various biochemical species such as ions [11–14], charged molecules (DNA (deoxyribonucleic acid) [15–17], protein biomarkers [18–22], polyelectrolytes [23,24]), virus-like particles [25,26], ligand-stabilized nanoparticles [27,28], etc. In addition, numerous enzyme-modified EISCAP biosensors were constructed for the detection of various analytes such as glucose [29–31], urea [30–32], creatinine [33], penicillin [31,34], formaldehyde [35], triglycerides [36], and acetoin [37]. The operation mechanism of these biosensors is based on the detection of local pH changes resulting from the catalytic reaction of the immobilized enzyme on the sensor surface with its specific substrate [10]. Moreover, the ability of EISCAP sensors for multi-analyte detection using a single EISCAP chip or an array of EISCAPs has been demonstrated [31,38–44].

Generally, the analytical characteristics of enzyme-based biosensors are strongly affected by the enzyme immobilization method [45,46]. Therefore, the choice of an appropriate enzyme immobilization strategy is a key factor for designing the biorecognition part of biosensors, including EISCAPs. The immobilization method must provide a high enzyme load on the biosensor surface, a good accessibility of the active sites for the target analytes, as well as to retain the structure, function, and catalytic activity of the enzyme [45,46]. Intensive efforts have been made during the last decade to develop novel immobilization techniques to improve the biosensor performance: the use of nanomaterials (e.g., metal, oxide or organic nanoparticles, nanowires, carbon nanotubes, magnetic beads) as nanoscale scaffolds for the immobilization of receptor molecules [47–50] as well as the incorporation of enzymes within nanoscale structures (e.g., alginate gels [51], polyelectrolyte/enzyme [52] or carbon nanotube/enzyme multilayers [53]). More recently, due to their regular shape, high surface-to-volume ratio, and extremely high density of surface docking sites, biological nanoscaffolds such as plant virus-like particles have increasingly been used for the precisely positioned immobilization of receptors on different transducers for biosensing purposes [54–56]. In contrast to various chemically synthesized nanoparticles, which are typically polydisperse with a more randomized and unreproducible distribution of particle size and density of shell molecules, many types of virus particles are monodisperse with regard to uniform morphologies and specific dimensions.

The *tobacco mosaic virus* (TMV) is one of the most comprehensively investigated plant viruses. Native full-length TMV particles are 300 nm long, nanotube-like nucleoprotein complexes with inner and outer diameters of 4 and 18 nm, respectively [56,57]. TMV particles exhibit excellent chemical and physical stability: in solutions, they can withstand 90 °C [58] and pH values between pH 3 and pH 9 [57]. The outer surface of each TMV nanotube holds thousands of docking sites accessible on about 2130 identical, helically arranged coat protein (CP) subunits, capable in the coupling of functional molecules. Therefore, TMV particles have often been utilized as biological nanoscaffolds for an extremely dense and precisely controlled immobilization of biorecognition molecules (receptors), including enzymes [55,56,59]. Moreover, TMV particles functionalized with bioreceptors can be simply combined with different transducer structures for chemical and biological sensing. For example, TMV particles were applied to assist the detection of the explosive agent trinitrotoluene [60], volatile organic compounds [61], and antigen-antibody binding [62,63]. The authors introduced amperometric glucose [64] as well as colorimetric and field-effect penicillin biosensors [65,66] by using TMV particles as nanoscaffolds for the immobilization of glucose oxidase and penicillinase, respectively. TMV particles as receptor nanocarriers usually enable increased receptor densities per sensor area, an enhanced mass transport of the target molecules to the TMV surface in comparison to planar surfaces, and a favorable orientation of the receptor molecules for an increased receptor-substrate interaction; all these factors enhance the overall biosensor performance [56,64].

In this work, the ability of TMV-assisted enzyme-based EISCAPs for multi-analyte detection was demonstrated. Penicillinase and urease were co-immobilized onto TMV surfaces for the sequential detection of penicillin and urea using the same p-Si-SiO$_2$-Ta$_2$O$_5$ field-effect sensor. These two enzyme/substrate systems (i.e., penicillinase/penicillin and urease/urea) represent typical model experiments inducing counter-rotating pH shifts on the sensor surface: a) generating hydrogen ions (penicillinase/penicillin → pH decrease) or b) consuming hydrogen ions (urease/urea → pH increase). For comparison, TMV-modified EISCAPs immobilized with either penicillinase or urease (no co-immobilization) were also studied. Finally, in a proof-of-concept experiment, the newly developed TMV-assisted bi-enzyme EISCAP biosensor was applied to mimic an XOR (Exclusive OR) enzyme logic gate.

2. Materials and Methods

2.1. Preparation of Biotinylated Tobacco Mosaic Virus Particles

A TMV variant (TMV$_{Cys}$) containing a S3C mutation close to the N-terminus of every CP was employed as a viral enzyme nanocarrier with a spacing of 2.5 to 3.5 nm of coupling sites on their outer protein coat [67]. This mutant exposed 2130 cysteine residues on each TMV particle presenting surficial sulfhydryl groups. TMV_{Cys} extracted from infected *Nicotiana tabacum* leaves were equipped with PEG_{11}-biotin moieties via maleimide-sulfhydryl-coupling (EZ-Link®Maleimide-PEG_{11}-Biotin, Thermo Scientific, Rockford, IL, USA) according to [56]. Briefly, a maleimide-PEG_{11}-biotin linker was incubated with TMV_{Cys} particles in a molar ratio of 3:1 (linker/CP) for 3 h at 26 °C under agitation. Unbound linker molecules were removed by centrifugal ultrafiltration (Amicon Ultra, 30 kDa molecular weight cut-off, Merck-Millipore, Darmstadt, Germany) in five consecutive washing steps with 10 mM of sodium-potassium-phosphate (SPP) buffer (pH 7.0). The resulting biotinylated TMV_{Cys} ($TMV_{Cys}/_{Bio}$) particles were resuspended and stored in this buffer at 4 °C. The particles were analyzed by sodium dodecyl sulphate–polyacrylamide gel electrophoresis and colloidal Coomassie Brilliant Blue G-250 staining [68]. Densitometric comparison of the signals corresponding to the biotinylated and the non-modified form of the CP confirmed biotinylation of >90 % of the CPs.

2.2. Preparation of Streptavidin-Enzyme Conjugates

Penicillinase from *Bacillus cereus* (1500–3000 Units/mg protein, Sigma Aldrich, Darmstadt, Germany) and urease from *Canavalia ensiformis* (75,265 Units/g solid, from Jack bean, Sigma Aldrich, Germany) were applied as the model enzymes. In order to immobilize them on the biotinylated TMV particles, they were conjugated with streptavidin (SA), allowing strong attachment via SA-biotin high affinity binding. The streptavidin conjugation was performed by utilizing a commercial conjugation kit (LYNX Streptavidin rapid conjugation kit, Bio-Rad, Great Britain). The conjugation of penicillinase was performed as described in [65], using a molar ratio between penicillinase and SA of 1:25 in the reaction mixture. For the urease conjugation, the same protocol was applied with a molar ratio of 1:10. The streptavidin-conjugated enzymes (SA-enzymes) were stored in stock solutions (10 mM phosphate buffered saline (PBS)) with a concentration of 600 Units/mL (SA-penicillinase) and of 2000 Units/mL (SA-urease) at 4 °C. For the fabrication of the bi-enzyme biosensors, both enzyme solutions were mixed in a ratio of 1:1, resulting in an enzyme cocktail containing 300 Units/mL SA-penicillinase and 1000 Units/mL SA-urease.

2.3. Modification of EISCAP Sensors with TMV Particles and Coupling of SA-Enzyme Conjugates

In this study, EISCAPs with an $Al/p-Si/SiO_2$ (30 nm)/Ta_2O_5 (60 nm) layered structure were used as the sensor platform, as schematically represented in Figure 1a. Details of the fabrication process were described in [25]. Prior to their modification with the TMV particles and enzymes, the sensors were cleaned in an ultrasonic bath for 5 min each with acetone, isopropanol, ethanol, and deionized water, and then installed into a homemade measurement cell. In the measurement cell, the sensor chip was sealed by an O-ring, with 0.5 cm^2 of the sensor surface in contact with the electrolyte (see Figure 1a). A 50 µL TMV solution (0.1 mg/mL) was incubated for one hour at room temperature (RT) on this exposed Ta_2O_5 surface to allow TMV adsorption. Subsequently, not-adsorbed TMV particles were washed away with 10 mM PBS buffer. Afterwards, 50 µL of the particular SA-enzyme solution (single enzyme EISCAP: 50 µL SA-penicillinase or 50 µL SA-urease solution; bi-enzyme EISCAP: 50 µL SA-penicillinase + SA-urease solution) were incubated for two hours at RT. The sensor surface was then flushed three times with 0.33 mM PBS and conditioned in 0.33 mM PBS buffer solution for at least one hour, before the electrochemical measurements were started. When the EISCAPs were not in use, they were stored in 0.33 mM PBS buffer at 4 °C.

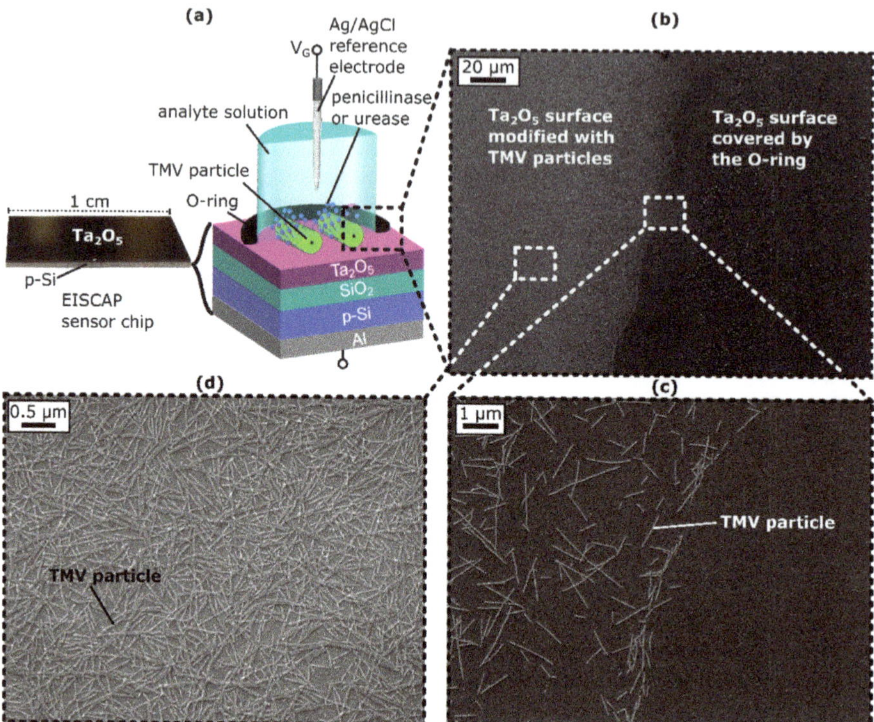

Figure 1. (**a**) Photo of the EISCAP sensor chip (left) and schematic layer structure (right) of the TMV-assisted Al/p-Si/SiO$_2$/Ta$_2$O$_5$-EISCAP sensor modified with penicillinase and/or urease, mounted in a measurement cell and sealed by an O-ring. (**b**) Scanning electron microscopic (SEM) image of the Ta$_2$O$_5$-gate surface showing the distinguished areas where the Ta$_2$O$_5$ surface is covered by the O-ring preventing TMV adsorption (right) and the area inside the O-ring, where the Ta$_2$O$_5$ is modified with TMV particles (left). (**c**) Magnification of the border line between TMV particle-modified and bare Ta$_2$O$_5$ surface. (**d**) Zoomed out image of the TMV particle-modified surface area. V_G: gate voltage.

2.4. Electrochemical Characterization of TMV-Modified EISCAP Biosensors

For electrochemical characterization of the TMV-assisted EISCAPs, an Ag/AgCl reference electrode (filled with 3 M KCl, Metrohm, Filderstadt, Germany) was immersed in the buffer solution and connected to an impedance analyzer (Zahner Zennium, Zahner Elektrik, Kronach, Germany). Furthermore, the Al rear-side contact was electrically connected to the impedance analyzer. All measurements were performed in a measurement buffer (0.33 mM PBS buffer) with varying concentrations of penicillin G (Sigma Aldrich, Darmstadt, Germany) and/or urea (GE Healthcare Bio-Sciences AB, Uppsala, Sweden) at RT. In order to avoid signal interferences, the set-up (except of the impedance analyzer) was integrated in a dark Faraday cage. As a first step, leakage-current measurements were carried out in a measurement solution without penicillin and urea. Therefore, gate voltages from –3 V to +3 V were applied with 100 mV steps between the reference electrode and the Al rear-side contact. These measurements served as a quality control of the insulator layer, where only sensors exhibiting a leakage current < 10 nA were selected for further electrochemical measurements. In the next step, capacitance–voltage (*C–V*) measurements were performed in the measurement buffer (by applying a gate voltage between –2 V and +2 V with 100 mV steps) to check the correct functioning of the field-effect EISCAPs. In order to measure the capacitance of the EISCAP, a small AC (alternating current) voltage of 20 mV with a frequency of 120 Hz was superimposed. For the following constant-capacitance

(ConCap) measurements, a working point was set in the depletion region of the C–V curve at about 60% of the maximum capacitance. The ConCap mode offered the time-dependent detection of surface-charge (potential) changes induced by, e.g., local pH changes at the sensor surface. During the ConCap mode, the capacitance of the sensor structure in the working point was kept constant by a control loop: Changes of the surface potential, e.g., induced by the enzymatic conversion of penicillin or urea, were compensated by applying an opposed voltage at the reference electrode. These voltage changes were recorded over time, allowing for the dynamic detection of changes in the penicillin and urea concentrations. A detailed description of the C–V and ConCap-measurement mode was provided in [23]. The ConCap measurements were conducted in a measurement buffer as well as in penicillin and urea solutions with concentrations between 0.1 and 5 mM for single-enzyme sensors and between 0.1 mM and 3 mM for bi-enzyme sensors. The penicillin and urea were purchased from Sigma Aldrich (Darmstadt, Germany) and GE Healthcare Bio-Sciences AB (Uppsala, Sweden), respectively.

2.5. Characterization of Surface Morphology by SEM

To control and characterize the TMV adsorption on the Ta_2O_5-gate surface, SEM images of the TMV-modified sensor chip were taken using a JEOL JSM-7800F Schottky field-emission microscope (JEOL GmbH, Freising, Germany). For this purpose, the sensors were mounted out of the measurement chamber, rinsed with deionized water, and dried with nitrogen to remove the salt residues of the buffer solution. Subsequently, an approximately 5 nm thick platinum-palladium layer was sputtered onto the sensor surface to provide conductivity and prevent the additional charging of the sensor surface.

3. Results and Discussion

3.1. SEM Images of TMV-Modified EISCAPs

To ensure that the TMV particles had adsorbed to the Ta_2O_5-gate surface within the O-ring and thus could act as nanocarriers for enzyme attachment, SEM images of the sensor surface were obtained after TMV loading. Figure 1b–d shows exemplary SEM images of the chip surface with a clear boundary between the TMV-modified region and the area sealed by the O-ring (see Figure 1c). Within the O-ring, TMV particles were homogeneously distributed on the Ta_2O_5 surface, with some virus particles present as lateral or head-to-tail aggregates (Figure 1d), as is typical for TMV particles [25,26,66]. The TMV particles were adsorbed in a high density of about 6.3×10^9 particles/cm^2 on the Ta_2O_5 surface, which was slightly higher than was reported in previous works [25,26,66] and revealed that the TMVs had been successfully attached to the Ta_2O_5 surface where they were available for enzyme immobilization.

3.2. TMV-Assisted Single-Enzyme EISCAPs

In order to study the sensor performance of TMV-assisted single enzyme EISCAPs, penicillin and urea biosensors with penicillinase and urease, respectively, were fabricated separately. Penicillinase catalyzes the conversion of penicillin to penicilloic acid, whereby H^+ ions are produced, leading to a local pH decrease [69]. In contrast, during the hydrolysis reaction of urea catalyzed by the urease, OH^- ions are produced (or H^+ ions are consumed), resulting in a local pH increase [70,71]. Changes in pH at the EISCAP surface cause the surface to become more positively charged (in the case of a pH decrease) or more negatively charged (in the case of a pH increase). This surface-charge change influences the width of the space-charge region in the semiconductor layer and, thus, the total capacitance of the sensor structure. Choosing these two enzymes enabled investigating whether EISCAPs modified with TMV particles as enzyme nanocarriers were suitable for enzymatic reactions involving both acidification and alkalization. The direction of the EISCAP-signal changes would directly correlate with these two kinds of enzymatic reactions, generating (penicillinase/penicillin) or consuming (urease/urea) hydrogen ions.

3.2.1. Penicillin Biosensor

Figure 2a shows a ConCap curve of the TMV/SA-penicillinase modified EISCAP biosensor in the loop of penicillin concentrations of 0.1, 0.5, 1, 3, 5, 3, 1, 0.5, and 0.1 mM. The measurements were carried out in 0.33 mM PBS at pH 8.0, which corresponds to the pH optimum of the penicillinase [72]. For each penicillin concentration, the ConCap signal was recorded for approximately 5 min.

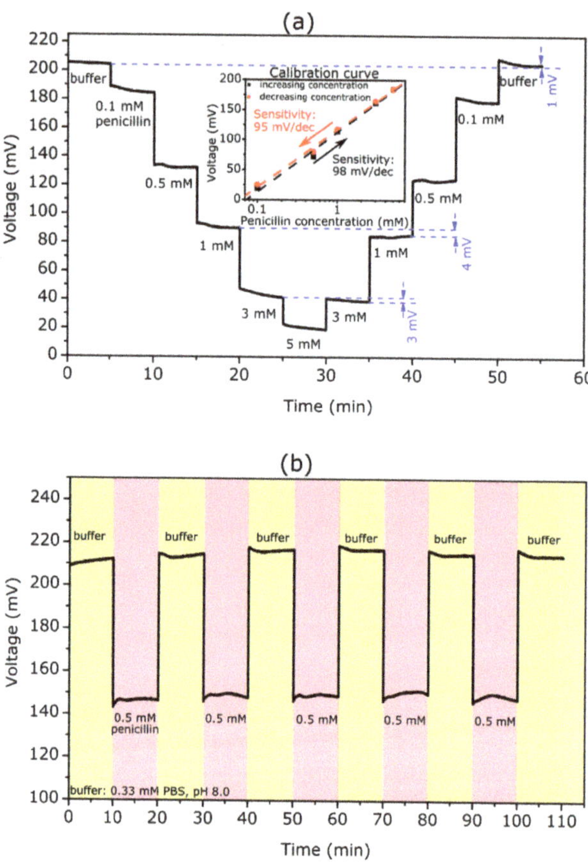

Figure 2. (**a**) ConCap curve of a TMV/SA-penicillinase-modified EISCAP recorded in 0.33 mM PBS buffer (pH 8.0) with different penicillin concentrations between 0.1 and 5 mM. The inset figure presents the resulting calibration curves with a penicillin sensitivity of about 98 and 95 mV/dec for the increasing (black) and decreasing (red) concentration series of measurements, respectively. (**b**) Reproducibility of the TMV-assisted EISCAP penicillin biosensor: the ConCap signal was repeatedly measured in buffer and in 0.5 mM penicillin solution in alternating order.

As expected, with increasing penicillin concentration, the recorded signal shifted towards less positive voltages. By increasing the penicillin concentration, more H^+ ions were generated and the local pH value at the sensor surface became lower: The produced H^+ ions protonate the hydroxyl groups on the Ta_2O_5 surface making it more positively charged. Consequently, the total capacitance of the EISCAP was decreased. To keep the total capacitance of the sensor constant, the applied voltage at the reference electrode should become more negative (or less positive), which is visible as a shift in the ConCap signal. Conversely, with decreasing penicillin concentration, the ConCap signal shifted in the opposite direction. The clear steps at different penicillin concentrations also underlined a fast response time of the TMV-based penicillin biosensor.

The measurement curve revealed that the sensor had a low hysteresis of 1 and 3 mV in the buffer without penicillin and in the 3 mM penicillin solution, respectively. However, it was observed that hysteresis is somewhat higher at lower penicillin concentrations with a maximum value of 8 mV at 0.5 mM. Generally, hysteresis of pH-sensitive field-effect devices is interpreted as slow response due to the slow buried sites underneath the gate-insulator surface (see e.g., [73,74]). It was reported that the hysteresis increases with an increasing pH-loop time, and in acid solutions it is smaller than in alkaline solutions [73,74] In addition, the hysteresis width may be affected by the background long-term drift, slow states at the Si-SiO$_2$ interface, as well as by the possible alteration of the enzyme activity due to local pH changes. The experiments performed in this study did not allow us to provide a clear explanation for the analyte-concentration dependence of the hysteresis width. Therefore, additional in-depth studies are needed to quantify this phenomenon.

The inset figure indicated the calibration plots for upward and downward penicillin-concentration loops evaluated from the ConCap response. The TMV-assisted EISCAP biosensor exhibited a high penicillin sensitivity of about 98 and 95 mV/dec for the increasing and decreasing concentration series of measurements, respectively. The small difference in penicillin sensitivities of 3 mV/dec observed for upward and downward concentration loops could be attributed to the hysteresis effect. The obtained sensitivity values were comparable to our previous work with TMV-based penicillin biosensors [66] and were higher than sensitivity values obtained for EISCAPs with adsorptively immobilized penicillinase (68.7 mV/dec) [31].

One of the important operation characteristics of biosensors is the reproducibility of the sensor response. To demonstrate the reproducibility of the TMV-assisted EISCAP penicillin biosensor, the ConCap signal was repeatedly measured in buffer (six times) and in a 0.5 mM penicillin solution (five times) in alternating order. The results of these experiments are shown in Figure 2b. The mean signal was 67 ± 1 mV, which underlines the high reproducibility of the developed penicillin biosensor.

3.2.2. Urea Biosensor

Figure 3a depicts a ConCap curve of an EISCAP modified with TMV/SA-urease, which was recorded in 0.33 mM PBS at pH 7.4, specified by the supplier as the optimum pH for urease [75].

Here, the ConCap signal behaved in the opposite way to that of the penicillin biosensor: with an increasing urea concentration from 0.1 to 5 mM, the measurement signal rose in the direction of the more positive (or less negative) voltages. The observed signal behavior could be explained as follows: an increase of the urea concentration resulted in a rise of the local pH value and deprotonation of the hydroxyl groups on the Ta$_2$O$_5$ surface, making it more negatively charged, whereby the total capacitance of the EISCAP increased. To keep the overall capacitance of the EISCAP sensor constant, the applied voltage at the reference electrode must be more positive (or less negative), which appeared as a signal shift in the ConCap curve.

Again, as with the penicillin measurements, clearly delineated signal levels could be seen at different urea concentrations with a low hysteresis of 2 mV for 3 mM urea. However, the hysteresis width increased with a decreasing urea concentration and amounts of 11 mV at 0.1 mM. In contrast to the penicillin measurements, a relatively large hysteresis of 8 mV was also recorded in the buffer solution. Additionally, at low urea concentrations, it took longer before a stable sensor signal was achieved. From the ConCap response at different urea concentrations, the calibration curves for the upward and downward concentration loops were evaluated, which are depicted in the inset figure. The TMV-assisted urea biosensor reveals a high urea sensitivity of 55 and 49 mV/dec for the increasing and decreasing concentration series of measurements. The difference in urea sensitivities for the upward and downward concentration loops was larger (6 mV/dec) than that of the

TMV-assisted penicillin biosensor. In other studies with urea sensors based on EISCAPs, urea sensitivities of 16 mV/dec (1 to 100 mM) by using a layer-by-layer nanofilm of ZnO nanocrystals and carbon nanotubes [76], 32 mV/dec by the immobilization of urease on magnetic particles [71], and 40.5 mV/dec (1 to 25 mM) in the case of nano-spotted urease [31] were achieved. Thus, the EISCAP modified with TMV/SA-urease offers a higher sensitivity and the ability of detecting low urea concentrations.

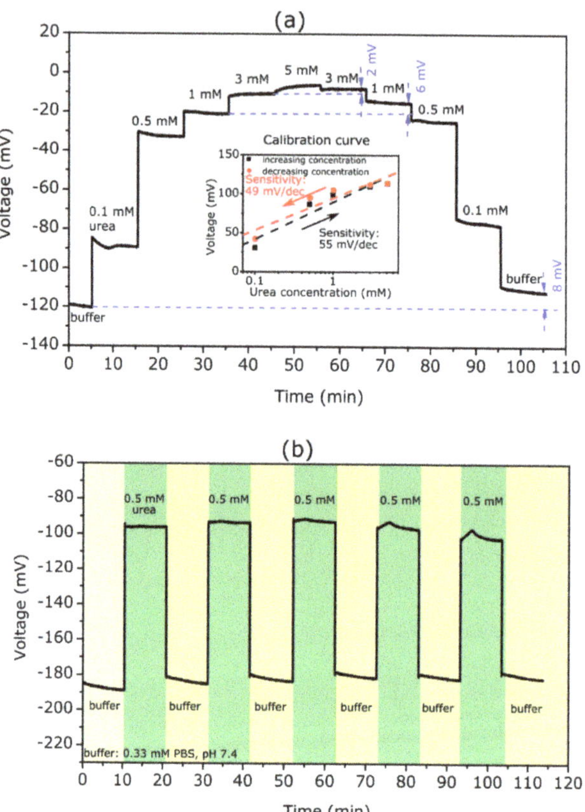

Figure 3. (**a**) ConCap curve of a TMV/SA-urease-modified EISCAP recorded in 0.33 mM PBS buffer (pH 7.4) with different urea concentrations between 0.1 and 5 mM. The inset figure presents the resulting calibration curves with a urea sensitivity of about 55 and 49 mV/dec for increasing (black) and decreasing (red) concentration series of measurements, respectively. (**b**) Reproducibility of the TMV-assisted EISCAP urea biosensor: the ConCap signal was repeatedly measured in buffer solution and in 0.5 mM urea solution in alternating order.

The results of the reproducibility experiments are shown in Figure 3b. Like the penicillin biosensor, the ConCap signal of the TMV-assisted EISCAP urea biosensor was repeatedly measured in buffer (six times) and in a 0.5 mM urea solution (five times) in alternating order. The urea biosensor signal was reproducible with a mean signal of (88 ± 5) mV.

3.3. TMV-Assisted Bi-Enzyme EISCAP Biosensor

The TMV-assisted bi-enzyme EISCAP, where both enzymes had been co-immobilized on the same sensor chip, was applied for the serial detection of urea and penicillin for the first time. Figure 4a depicts the ConCap response at different urea and penicillin concentrations ranging from 0.1 to 3 mM.

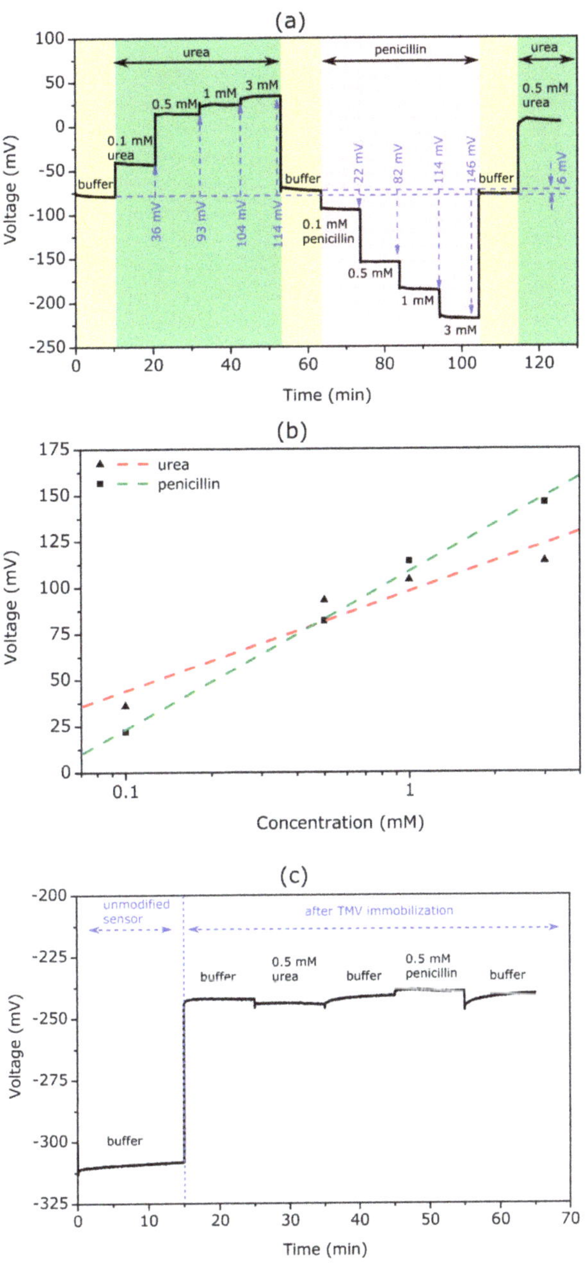

Figure 4. (**a**) ConCap curve of the TMV-assisted bi-enzyme EISCAP biosensor recorded in buffer and at different concentrations of urea and penicillin. (**b**) Calibration curves for urea and penicillin evaluated from the ConCap response in (**a**). (**c**) ConCap curve recorded with an EISCAP sensor before and after TMV loading (without enzymes) in 0.33 mM PBS (pH 7.4) and in 0.33 mM PBS (pH 7.4) with 0.5 mM urea and 0.5 mM penicillin, respectively.

The measurements were performed in the following order: First, the ConCap response was recorded in the buffer solution (0.33 mM PBS, pH 7.4) to obtain the baseline signal for the sensor, followed by measurements in urea solutions with different concentrations. Second, the sensor signal was recorded again in the buffer solution to obtain the baseline signal for the subsequent penicillin measurements with varying penicillin concentrations. Finally, the ConCap signal was recorded again in the buffer and urea solution. Such sequential arrangements of experiments are useful to examine the recoverability of the sensor signal as well as to identify possible hysteresis and drift effects.

As expected, the bi-enzyme sensor showed distinct signal steps towards more positive voltages with an increasing urea concentration and towards more negative voltages with an increasing penicillin concentration. For example, at a urea and penicillin concentration of 3 mM, the signal shift reached approximately 114 and 146 mV, respectively. Taking into account a pH sensitivity of 56 mV/pH for the Ta_2O_5-gate EISCAPs studied in this work, the local pH value on the TMV/SA-enzyme modified EISCAP surface was estimated to be pH = 9.4 or pH = 4.8 at a urea or penicillin concentration of 3 mM, respectively. This indicated the stability of the TMV/SA-enzyme system on the EISCAP surface over a wide pH range.

In Figure 4b, the evaluated calibration curves of the TMV-assisted bi-enzyme EIS-CAP for urea and penicillin are illustrated. The biosensor exhibited a urea sensitivity of 54 mV/dec, which was nearly similar to that of the single-enzyme biosensor. The penicillin sensitivity of the bi-enzyme EISCAP amounted to 85 mV/dec, which was slightly lower than that achieved with the single-enzyme EISCAP (see Section 3.2.1). The results indicated that both enzymes maintained their activity after the co-immobilization.

In general, a possible non-specific adsorption of analyte molecules onto the gate surface of the field-effect device may induce an unwanted background signal, and, thus, may reduce the signal-to-noise ratio of the sensor signal [1,4,10]. In our case, such non-specific adsorption may occur either on the TMV particles or on TMV-free areas of the gate surface of the EISCAP. Therefore, we studied the signal behavior of the TMV-modified (but enzyme-free) EISCAP in the urea and penicillin solutions. Figure 4c depicts the ConCap signal of the EISCAP recorded in 0.33 mM PBS (pH 7.4) buffer before (unmodified sensor) and after the loading of TMV particles, followed by measurements in buffer containing 0.5 mM urea or penicillin. As TMV particles are negatively charged [26], the sensor signal shifted to more positive voltages after their loading onto the Ta_2O_5 surface. However, upon subsequent measurements in analyte solutions, the signal remained almost constant, indicating that the existence of urea or penicillin molecules in the solution had practically no impact on the EISCAP sensor signal.

3.4. XOR Logic Gate Using TMV-Assisted Bi-Enzyme Biosensor

In the past, intensive research was performed in the field of enzyme-based logic gates, which mimic the operation of electronic logic gates (see e.g., [77,78]). In enzyme logic gates, Boolean logic operations are activated by specific molecular inputs via enzymatic reactions. An interfacing of enzyme logic gates with electronic transducers was considered a very promising approach for designing digital biosensors that could provide qualitative evidence (in a YES/NO format) concerning the presence or absence of a specific analyte in the sample [79–81]. In previous works, we demonstrated the successful integration of enzyme logic gates with pH-sensitive EISCAPs as **AND**-Reset, **OR**-Reset, **CNOT** (**controlled NOT**) and **XOR** gates [82–85]. In these logic gate devices, enzymes were immobilized onto the biosensor surface by means of physical entrapment within a membrane or through physical adsorption. In this work, the EISCAP modified with TMV particles (as bi-enzyme nanocarriers) was applied to mimic an **XOR** enzyme logic gate.

The **XOR** gate is one of the important elements of biocomputing systems: it provides a true output (**1**) if only one of the inputs is true. If both input signals are false (**0**) or both are true (**1**), it must remain inactive (output signal **0**). As was noted in [86], the enzyme-based **XOR** gate is difficult to realize, because it requires enzymatic reactions that are able to produce approximately equal signals of opposite direction (net output signal should be 0), when both input signals simultaneously appear. The enzyme/substrate systems used in this work (i.e., urease/urea and penicillinase/penicillin) represented typical examples of such enzymatic reactions producing pH changes and consequently EISCAP signals in opposite direction (see Figure 4).

Figure 5a shows the schematic symbol and truth table for the enzyme-based **XOR** gate with urea as input A and penicillin as input B. The expected ConCap signal of the TMV-assisted bi-enzyme EISCAP biosensor for different input combinations is illustrated in Figure 5b.

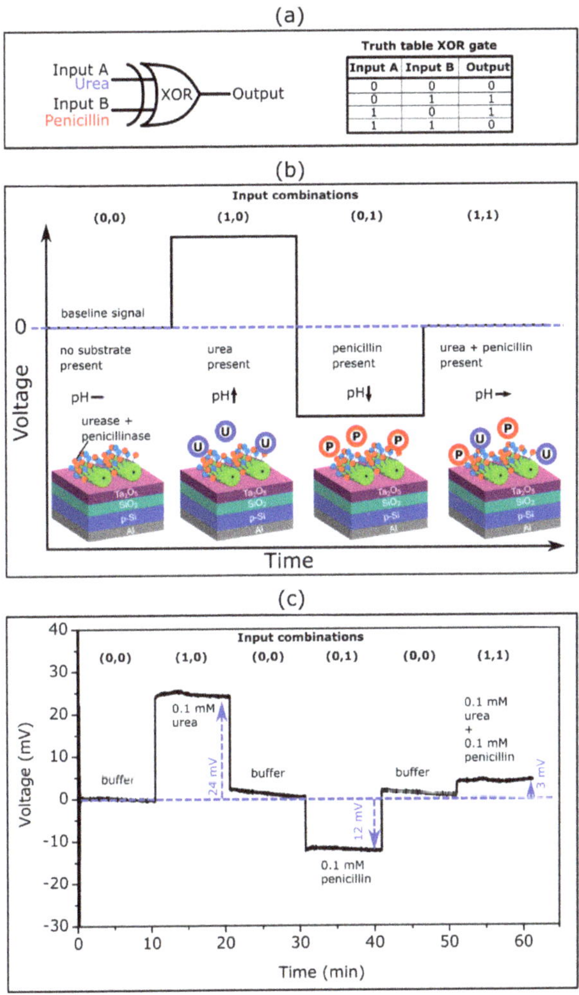

Figure 5. (a) Schematic symbol and truth table of an **XOR** gate with urea and penicillin as input signals. (b) Expected ConCap response for different input combinations. (c) Experimental ConCap curve of the TMV-assisted bi-enzyme EISCAP biosensor recorded in buffer (input combination (0,0)), in 0.1 mM urea or penicillin solution (input combination (1,0) or (0,1)) and in a mixture of 0.1 mM urea and 0.1 mM penicillin solution (input combination (1,1)). U: urea, P: penicillin.

The presence of the respective analyte in the solution corresponded to the input signal **1**, while the absence of an analyte was considered as the input signal **0**. When no analyte was present in the solution (input combination **(0,0)**), the resulting output signal became **0**, because no local pH change occurred. If one of the analytes (urea or penicillin) was present (input combination **(1,0)** or **(0,1)**), the output signal became **1** due to the enzymatically induced pH change. The presence of both analytes led to an output signal **0**, as both enzymatic reactions occurred simultaneously with opposing pH shifts which compensated each other.

Figure 5c represents an experimental ConCap logic signal of the TMV-assisted bi-enzyme EISCAP biosensor corresponding to various input combinations: The measurement was started in the buffer solution in the absence of the analyte (input combination **(0,0)**), followed by measurement in a 0.1 mM urea (input combination **(1,0)**) and 0.1 mM penicillin solution (input combination **(0,1)**). To return the local pH to its initial value, after each measurement in the particular analyte solution, the biosensor was exposed to the buffer solution and the baseline signal was recorded again. Finally, the ConCap signal was recorded in the solution containing both 0.1 mM urea and 0.1 mM penicillin (input combination **(1,1)**). Relatively large signal shifts of opposite direction were observed when only one of the analytes was present in the solution (24 mV at 0.1 mM urea; 12 mV at 0.1 mM penicillin). In contrast, a very small signal shift of 3 mV was detected if both analytes were present in the solution: two simultaneously occurring enzymatic reactions produced opposing pH changes, resulting in a nearly negligible sensor output signal.

The described proof-of-principle experiment demonstrated the successful application of the TMV-assisted bi-enzyme EISCAPs for the development of **XOR** enzyme logic gates. Future work will be directed to minimize the output signal (should be ideally **0**) of the EISCAP in the presence of both analytes via the optimization of the ratio of urease/penicillinase activities and/or urea/penicillin concentrations in the solution.

4. Conclusions

The appropriate enzyme immobilization strategy is a key factor for the design of enzyme-based biosensors. Plant virus-like particles offer the possibility of extremely dense and precise immobilization of enzymes, due to uniform 3D structures and a high density of surface binding sites. In this study, TMV particles were utilized for the co-immobilization of penicillinase and urease onto the Ta_2O_5-gate surface of an EISCAP sensor for the serial detection of penicillin and urea. These two enzyme/substrate systems (i.e., penicillinase/penicillin and urease/urea) allowed the performance of typical model experiments with opposing pH shifts on the sensor surface: (a) generating hydrogen ions (penicillinase/penicillin with associated pH decrease) or (b) consuming hydrogen ions (urease/urea with associated pH increase). The sensitive characteristics of TMV-assisted bi-enzyme EISCAPs were characterized in the ConCap mode and compared with those of single-enzyme EISCAPs.

The single-enzyme biosensors offered a high penicillin (95–98 mV/dec) and urea (49–55 mV/dec) sensitivity and remarkable reproducibility. These notable sensor properties were retained even when both enzymes were co-immobilized on the TMV particles, resulting in a penicillin and urea sensitivity of 85 and 54 mV/dec, respectively. The successful application of the TMV-assisted biosensor for designing an **XOR** enzymatic gate further highlighted the potential of the presented sensor arrangement.

The results achieved in this study demonstrated the great prospects of TMV particles as enzyme nanocarriers in constructing EISCAP biosensors for multi-analyte detection. In future studies, TMV-assisted multi-enzyme EISCAPs could be extended to other enzymes. Furthermore, TMV-assisted EISCAPs could be employed for the characterization and application of two- or multi-step enzymatic cascades. Due to the uniform 3D structure of the TMV particles, the enzymes are geometrically close to each other, whereby enzymatic product/substrate diffusion can optimally take place, and the influence of interfering substances could be reduced. This may include not only arrangements with blends of distinct enzymes immobilized on the TMV nanoscaffolds but could also be tailored to enzyme systems collaborating between separate particles. If installed on TMV carriers with length-defined, selectively addressable longitudinal subdomains, such sensors might even provide fundamental insights into spacing-dependent interactions of enzyme groups and, thus, ease the design of high-efficiency artificial biocatalytic systems [87].

Author Contributions: Conceptualization, M.W., A.P., M.K. and M.J.S.; methodology, M.W., A.P., F.V., T.W., C.W. and M.J.S.; validation, M.W., A.P. and M.J.S.; formal analysis, M.W., A.P., F.V. and M.J.S.; investigation, M.W., F.V. and T.W.; writing, M.W., A.P., T.W., M.K., C.W. and M.J.S.; supervision, M.K. and M.J.S. All authors have read and agreed to the published version of the manuscript.

Funding: This work was funded by the Deutsche Forschungsgemeinschaft (DFG: German Research Foundation)—446507449.

Institutional Review Board Statement: Not applicable.

Informed Consent Statement: Not applicable.

Data Availability Statement: The data presented in this study are available on request from the corresponding author.

Acknowledgments: The authors gratefully thank H. Iken, D. Rolka, T. Karschuck, and D. Rashid for technical support.

Conflicts of Interest: The authors declare no conflict of interest. The funders had no role in the design of the study; in the collection, analyses, or interpretation of data; in the writing of the manuscript; or in the decision to publish the results.

References

1. Poghossian, A.; Schöning, M.J. Recent progress in silicon-based biologically sensitive field-effect devices. *Curr. Opin. Electrochem.* **2021**, *29*, 100811. [CrossRef]
2. Gao, A.; Chen, S.; Wang, Y.; Li, T. Silicon nanowire field-effect-transistor-based biosensor for biomedical applications. *Sens. Mater.* **2018**, *30*, 1619–1628. [CrossRef]
3. Syu, Y.-C.; Hsu, W.-E.; Lin, C.-T. Field-effect transistor biosensing: Devices and clinical applications. *ECS J. Solid State Sci. Technol.* **2018**, *7*, Q3196–Q3207. [CrossRef]
4. Poghossian, A.; Jablonski, M.; Molinnus, D.; Wege, C.; Schöning, M.J. Field-effect sensors for virus detection: From Ebola to SARS-CoV-2 and plant viral enhancers. *Front. Plant. Sci.* **2020**, *11*, 598103. [CrossRef] [PubMed]
5. Pullano, S.A.; Critello, C.D.; Mahbub, I.; Tasneem, N.T.; Shamsir, S.; Islam, S.K.; Greco, M.; Fiorillo, A.S. EGFET-based sensors for bioanalytical applications: A review. *Sensors* **2018**, *18*, 4042. [CrossRef]
6. De Moraes, A.C.M.; Kubota, L.T. Recent trends in field-effect transistors-based immunosensors. *Chemosensors* **2016**, *4*, 20. [CrossRef]
7. Sakata, T. Biologically coupled gate field-effect transistors meet in vitro diagnostics. *ACS Omega* **2019**, *4*, 11852–11862. [CrossRef]
8. Yoshinobu, T.; Miyamoto, K.; Werner, C.F.; Poghossian, A.; Wagner, T.; Schöning, M.J. Light-addressable potentiometric sensors for quantitative spatial imaging of chemical species. *Annu. Rev. Anal. Chem.* **2017**, *10*, 225–246. [CrossRef]
9. Yoshinobu, T.; Schöning, M.J. Light-addressable potentiometric sensors for cell monitoring and biosensing. *Curr. Opin. Electrochem.* **2021**, *28*, 100727. [CrossRef]
10. Poghossian, A.; Schöning, M.J. Capacitive field-effect chemical sensors and biosensors: A status report. *Sensors* **2020**, *20*, 5639. [CrossRef]
11. Chen, M.; Jin, Y.; Qu, X.; Jin, W.; Zhao, J. Electrochemical impedance spectroscopy study of Ta_2O_5 based EIOS pH sensors in acid environment. *Sens. Actuators B* **2014**, *192*, 399–405. [CrossRef]
12. Molinnus, D.; Iken, H.; Johnen, A.L.; Richstein, B.; Hellmich, L.; Poghossian, A.; Knoch, J.; Schöning, M.J. Miniaturized pH-sensitive field-effect capacitors with ultrathin Ta_2O_5 films prepared by atomic layer deposition. *Phys. Status Solidi A* **2021**. submitted.
13. Ye, Y.-R.; Wang, J.-C.; Chan, Y.-T. Anion sensing and interfering behaviors of electrolyte–insulator–semiconductor sensors with nitrogen plasma-treated samarium oxide. *Jpn. J. Appl. Phys.* **2015**, *54*, 04DL04. [CrossRef]
14. Cho, H.; Kim, K.; Meyyappan, M.; Baek, C.-K. LaF_3 electrolyte-insulator-semiconductor sensor for detecting fluoride ions. *Sens. Actuators B* **2019**, *279*, 183–188. [CrossRef]
15. Pan, T.-M.; Chang, K.-Y.; Lin, C.-W.; Tsai, S.-W.; Wu, M.H. Label-free detection of DNA using high-k $Lu_2Ti_2O_7$ electrolyte-insulator-semiconductors. *J. Mater. Chem.* **2012**, *22*, 1358–1363. [CrossRef]
16. Bronder, T.S.; Jessing, M.P.; Poghossian, A.; Keusgen, M.; Schöning, M.J. Detection of PCR-amplified tuberculosis DNA fragments with polyelectrolyte-modified field-effect sensors. *Anal. Chem.* **2018**, *90*, 7747–7753. [CrossRef]

17. Branquinhoa, R.; Veigas, B.; Pinto, J.V.; Martins, R.; Fortunato, E.; Baptista, P.V. Real-time monitoring of PCR amplification of proto-oncogene c-MYC using a Ta_2O_5 electrolyte–insulator–semiconductor sensor. *Biosens. Bioelectron.* **2011**, *28*, 44–49. [CrossRef] [PubMed]

18. Kumar, N.; Kumar, S.; Kumar, J.; Panda, S. Investigation of mechanisms involved in the enhanced label free detection of prostate cancer biomarkers using field effect devices. *J. Electrochem. Soc.* **2017**, *164*, B409–B416. [CrossRef]

19. Chand, R.; Han, D.; Neethirajan, S.; Kim, Y.-S. Detection of protein kinase using an aptamer on a microchip integrated electrolyte-insulator- semiconductor sensor. *Sens. Actuators B* **2017**, *248*, 973–979. [CrossRef]

20. Pan, T.-M.; Lin, T.-W.; Chen, C.-Y. Label-free detection of rheumatoid factor using YbY_xO_y electrolyte-insulator-semiconductor devices. *Anal. Chim. Acta* **2015**, *891*, 304–311. [CrossRef] [PubMed]

21. Hlukhova, H.; Menger, M.; Offenhäusser, A.; Vitusevich, S. Highly sensitive aptamer-based method for the detection of cardiac biomolecules on silicon dioxide surfaces. *MRS. Adv.* **2016**, *3*, 1535–1541. [CrossRef]

22. Bahri, M.; Baraket, A.; Zine, N.; Ali, M.B.; Bausells, J.; Errachid, A. Capacitance electrochemical biosensor based on silicon nitride transducer for TNF-α cytokine detection in artificial human saliva: Heart failure (HF). *Talanta* **2020**, *209*, 120501. [CrossRef] [PubMed]

23. Poghossian, A.; Weil, M.; Cherstvy, A.G.; Schöning, M.J. Electrical monitoring of polyelectrolyte multilayer formation by means of capacitive field-effect devices. *Anal. Bioanal. Chem.* **2013**, *405*, 6425–6436. [CrossRef] [PubMed]

24. Garyfallou, G.Z.; de Smet, L.C.P.M.; Sudhölter, E.J.R. The effect of the type of doping on the electrical characteristics of electrolyte-oxide-silicon sensors: pH sensing and polyelectrolyte adsorption. *Sens. Actuators B* **2012**, *168*, 207–213. [CrossRef]

25. Jablonski, M.; Poghossian, A.; Severins, R.; Keusgen, M.; Wege, C.; Schöning, M.J. Capacitive field-effect biosensor studying adsorption of *tobacco mosaic virus* particles. *Micromachines* **2021**, *12*, 57. [CrossRef]

26. Jablonski, M.; Poghossian, A.; Keusgen, M.; Wege, C.; Schöning, M.J. Detection of plant virus particles with a capacitive field-effect sensor. *Anal. Bioanal. Chem.* **2021**, *413*, 5669–5678. [CrossRef]

27. Gun, J.; Rizkov, D.; Lev, O.; Abouzar, M.H.; Poghossian, A.; Schöning, M.J. Oxygen plasma-treated gold nanoparticle-based field-effect devices as transducer structures for bio-chemical sensing. *Microchim. Acta* **2009**, *164*, 395–404. [CrossRef]

28. Karschuck, T.; Kaulen, C.; Poghossian, A.; Wagner, P.H.; Schöning, M.J. Gold nanoparticle-modified capacitive field-effect sensors: Studying the surface density of nanoparticles and coupling of charged polyelectrolyte macromolecules. *Electrochem. Sci. Adv.* **2021**, e2100179. [CrossRef]

29. Lin, Y.-H.; Das, A.; Wu, M.-H.; Pan, T.-M.; Lai, C.-S. Microfluidic chip integrated with an electrolyte-insulator-semiconductor sensor for pH and glucose level measurement. *Int. J. Electrochem. Sci.* **2013**, *8*, 5886–5901.

30. Lin, C.F.; Kao, C.H.; Lin, C.Y.; Chen, K.L.; Lin, Y.H. NH_3 plasma-treated magnesium doped zinc oxide in biomedical sensors with electrolyte-insulator-semiconductor (EIS) structure for urea and glucose applications. *Nanomaterials* **2020**, *10*, 583. [CrossRef]

31. Molinnus, D.; Beging, S.; Lowis, C.; Schöning, M.J. Towards a multi-enzyme capacitive field-effect biosensor by comparative study of drop-coating and nano-spotting technique. *Sensors* **2020**, *20*, 4924. [CrossRef] [PubMed]

32. Pan, T.-M.; Lin, C.-W. High-k Dy_2TiO_5 electrolyte-insulator-semiconductor urea biosensors. *J. Electrochem. Soc.* **2011**, *158*, J100–J105. [CrossRef]

33. Pan, T.-M.; Lin, C.-W.; Lin, W.-Y.; Wu, M.-H. High-k $Tm_2Ti_2O_7$ electrolyte-insulator semiconductor creatinine biosensor. *IEEE Sens. J.* **2011**, *11*, 2388–2394. [CrossRef]

34. Beging, S.; Leinhos, M.; Jablonski, M.; Poghossian, A.; Schöning, M.J. Studying the spatially resolved immobilization of enzymes on a capacitive field-effect structure by means of nano-spotting. *Phys. Status Solidi A* **2015**, *212*, 1353–1358. [CrossRef]

35. Ben Ali, M.; Gonchar, M.; Gayda, G.; Paryzhak, S.; Maaref, M.A.; Jaffrezic-Renault, N.; Korpan, Y. Formaldehyde-sensitive sensor based on recombinant formaldehyde dehydrogenase using capacitance versus voltage measurements. *Biosens. Bioelectron.* **2007**, *22*, 2790–2795. [CrossRef]

36. Mathew, A.; Pandian, G.; Bhattacharya, E.; Chadha, A. Novel applications of silicon and porous silicon based EISCAP biosensors. *Phys. Status Solidi A* **2009**, *206*, 1369–1373. [CrossRef]

37. Jablonski, M.; Münstermann, F.; Nork, J.; Molinnus, D.; Muschallik, L.; Bongaerts, J.; Wagner, T.; Keusgen, M.; Siegert, P.; Schöning, M.J. Capacitive field-effect biosensor applied for the detection of acetoin in alcoholic beverages and fermentation broths. *Phys. Status Solidi A* **2021**, *218*, 2000765. [CrossRef]

38. Taing, M.; Sweatman, D. Fabrication techniques for an arrayed EIS biosensor. In Proceedings of the Electronics Packaging Technology Conference (IEEE), Singapore, 9–11 December 2009; pp. 168–173. [CrossRef]

39. Abouzar, M.H.; Poghossian, A.; Pedraza, A.M.; Gandhi, D.; Ingebrandt, S.; Moritz, W.; Schöning, M.J. An array of field-effect nanoplate SOI capacitors for (bio-)chemical sensing. *Biosens. Bioelectron.* **2011**, *26*, 3023–3028. [CrossRef] [PubMed]

40. Poghossian, A.; Welden, R.; Buniatyan, V.V.; Schöning, M.J. An array of on-chip integrated, individually addressable capacitive field-effect sensors with control gate: Design and modelling. *Sensors* **2021**, *21*, 6161. [CrossRef]

41. Kao, C.H.; Chen, H.; Hou, F.Y.S.; Chang, S.W.; Chang, C.W.; Lai, C.S.; Chen, C.P.; He, Y.Y.; Lin, S.-R.; Hsieh, K.M.; et al. Fabrication of multianalyte CeO_2 nanograin electrolyte-insulator-semiconductor biosensors by using CF_4 plasma treatment. *Sens. Bio-Sens. Res.* **2015**, *5*, 71–77. [CrossRef]

42. Lin, Y.-H.; Wang, S.-H.; Wu, M.-H.; Pan, T.-M.; Lai, C.-S.; Luo, J.-D.; Chiou, C.-C. Integrating solid-state sensor and microfluidic devices for glucose, urea and creatinine detection based on enzyme-carrying alginate microbeads. *Biosens. Bioelectron.* **2013**, *43*, 328–335. [CrossRef]

43. Kao, C.H.; Chen, H.; Lee, M.L.; Liu, C.C.; Ueng, H.-Y.; Chu, Y.C.; Chen, Y.J.; Chang, K.M. Multianalyte biosensor based on pH-sensitive ZnO electrolyte–insulator–semiconductor structures. *J. Appl. Phys.* **2014**, *115*, 184701. [CrossRef]
44. Lin, Y.-H.; Chiang, C.-H.; Wu, M.-H.; Pan, T.-M.; Luo, J.-D.; Chiou, C.-C. Solid-state sensor incorporated in microfluidic chip and magnetic-bead enzyme immobilization approach for creatinine and glucose detection in serum. *Appl. Phys. Lett.* **2011**, *99*, 253704. [CrossRef]
45. Ansari, S.A.; Husain, Q. Potential applications of enzymes immobilized on/in nano materials: A review. *Biotechnol. Adv.* **2012**, *30*, 512–523. [CrossRef] [PubMed]
46. Sassolas, A.; Blum, L.J.; Leca-Bouvier, B.D. Immobilization strategies to develop enzymatic biosensors. *Biotechnol. Adv.* **2012**, *30*, 489–511. [CrossRef]
47. Poghossian, A.; Bäcker, M.; Mayer, D.; Schöning, M.J. Gating capacitive field-effect sensors by the charge of nanoparticle/molecule hybrids. *Nanoscale* **2015**, *7*, 1023–1031. [CrossRef] [PubMed]
48. Gun, J.; Schöning, M.J.; Abouzar, M.H.; Poghossian, A.; Katz, E. Field-effect nanoparticle-based glucose sensor on a chip: Amplification effect of coimmobilized redox species. *Electroanalysis* **2008**, *20*, 1748–1753. [CrossRef]
49. Braham, Y.; Barhoumi, H.; Maaref, A. Urease capacitive biosensors using functionalized magnetic nanoparticles for atrazine pesticide detection in environmental samples. *Anal. Methods* **2013**, *5*, 4898–4904. [CrossRef]
50. Kiralp, S.; Topcu, A.; Bayramoğlu, G.; Arıca, M.Y.; Toppare, L. Alcohol determination via covalent enzyme immobilization on magnetic beads. *Sens. Actuators B* **2008**, *128*, 521–528. [CrossRef]
51. Lin, Y.-H.; Chu, C.-P.; Lin, C.-F.; Liao, H.-H.; Tsai, H.-H.; Juang, Y.-Z. Extended-gate field-effect transistor packed in micro channel for glucose, urea and protein biomarker detection. *Biomed Microdevices* **2015**, *17*, 111. [CrossRef]
52. Abouzar, M.H.; Poghossian, A.; Siqueira, J.R., Jr.; Oliveira, O.N., Jr.; Moritz, W.; Schöning, M.J. Capacitive electrolyte-insulator-semiconductor structures functionalized with a polyelectrolyte/enzyme multilayer: New strategy for enhanced field-effect biosensing. *Phys. Status Solidi A* **2010**, *207*, 884–890. [CrossRef]
53. Siqueira, J.R., Jr.; Abouzar, M.H.; Poghossian, A.; Zucolotto, V.; Oliveira, O.N., Jr.; Schöning, M.J. Penicillin biosensor based on a capacitive field-effect structure functionalized with a dendrimer/carbon nanotube multilayer. *Biosens. Bioelectron.* **2009**, *25*, 497–501. [CrossRef] [PubMed]
54. Aljabali, A.A.A.; Barclay, J.E.; Steinmetz, N.F.; Lomonossoff, G.P.; Evans, D.J. Controlled immobilisation of active enzymes on the *cowpea mosaic virus* capsid. *Nanoscale* **2012**, *4*, 5640–5645. [CrossRef] [PubMed]
55. Cardinale, D.; Carette, N.; Michon, T. Virus scaffolds as enzyme nano-carriers. *Trends Biotechnol.* **2012**, *30*, 369–376. [CrossRef]
56. Koch, C.; Wabbel, K.; Eber, F.J.; Krolla-Sidenstein, P.; Azucena, C.; Gliemann, H.; Eiben, S.; Geiger, F.; Wege, C. Modified TMV particles as beneficial scaffolds to present sensor enzymes. *Front. Plant. Sci.* **2015**, *6*, 1137. [CrossRef] [PubMed]
57. Alonso, J.M.; Gorzny, M.Ł.; Bittner, A.M. The physics of *tobacco mosaic virus* and virus-based devices in biotechnology. *Trends Biotechnol.* **2013**, *31*, 530–538. [CrossRef]
58. Wege, C.; Koch, C. From stars to stripes: RNA-directed shaping of plant viral protein templates-structural synthetic virology for smart biohybrid nanostructures. *Wiley Interdiscip. Rev. Nanomed. Nanobiotechnol.* **2020**, *12*, e1591. [CrossRef]
59. Calò, A.; Eiben, S.; Okuda, M.; Bittner, A.M. Nanoscale device architectures derived from biological assemblies: The case of *tobacco mosaic virus* and (apo)ferritin. *Jpn. J. Appl. Phys.* **2016**, *55*, 03DA01. [CrossRef]
60. Zang, F.; Gerasopoulos, K.; Fan, X.Z.; Brown, A.D.; Culver, J.N.; Ghodssi, R. An electrochemical sensor for selective TNT sensing based on *tobacco mosaic virus*-like particle binding agents. *Chem. Commun.* **2014**, *50*, 12977–12980. [CrossRef]
61. Bruckman, M.A.; Liu, J.; Koley, G.; Li, Y.; Benicewicz, B.; Niu, Z.W.; Wang, Q.A. *Tobacco mosaic virus* based thin film sensor for detection of volatile organic compounds. *J. Mater. Chem.* **2010**, *20*, 5715–5719. [CrossRef]
62. Fan, X.Z.; Naves, L.; Siwak, N.P.; Brown, A.; Culver, J.; Ghodssi, R. Integration of genetically modified virus-like-particles with an optical resonator for selective bio-detection. *Nanotechnology* **2015**, *26*, 205501. [CrossRef] [PubMed]
63. Zang, F.; Gerasopoulos, K.; Brown, A.D.; Culver, J.N.; Ghodssi, R. Capillary microfluidics-assembled virus-like particle bionanoreceptor interfaces for label-free biosensing. *ACS Appl. Mater. Interfaces* **2017**, *9*, 8471–8479. [CrossRef]
64. Bäcker, M.; Koch, C.; Eiben, S.; Geiger, F.; Eber, F.; Gliemann, H.; Poghossian, A.; Wege, C.; Schöning, M.J. *Tobacco mosaic virus* as enzyme nanocarrier for electrochemical biosensors. *Sens. Actuators B* **2017**, *238*, 716–722. [CrossRef]
65. Koch, C.; Poghossian, A.; Schöning, M.J.; Wege, C. Penicillin detection by *tobacco mosaic virus*-assisted colorimetric biosensors. *Nanotheranostics* **2018**, *2*, 184–196. [CrossRef]
66. Poghossian, A.; Jablonski, M.; Koch, C.; Bronder, T.S.; Rolka, D.; Wege, C.; Schöning, M.J. Field-effect biosensor using virus particles as scaffolds for enzyme immobilization. *Biosens. Bioelectron.* **2018**, *110*, 168–174. [CrossRef]
67. Geiger, F.C.; Eber, F.J.; Eiben, S.; Mueller, A.; Jeske, H.; Spatz, J.P.; Wege, C. TMV nanorods with programmed longitudinal domains of differently addressable coat proteins. *Nanoscale* **2013**, *5*, 3808–3816. [CrossRef] [PubMed]
68. Dyballa, N.; Metzger, S. Fast and sensitive colloidal coomassie G-250 staining for proteins in polyacrylamide gels. *J. Vis. Exp.* **2009**, 30. [CrossRef]
69. Welden, R.; Jablonski, M.; Wege, C.; Keusgen, M.; Wagner, P.H.; Wagner, T.; Schöning, M.J. Light-addressable actuator-aensor platform for monitoring and manipulation of pH gradients in microfluidics: A case study with the enzyme penicillinase. *Biosensors* **2021**, *11*, 171. [CrossRef]
70. Rutherford, J.C. The emerging role of urease as a general microbial virulence factor. *PLoS Pathogens* **2014**, *10*, e1004062. [CrossRef] [PubMed]

71. Nouira, W.; Barhoumi, H.; Maaref, A.; Renault, N.J.; Siadat, M. Tailoring of analytical performances of urea biosensors using nanomaterials. *J. Phys.: Conf. Ser.* **2013**, *416*, 012010. [CrossRef]

72. Sawai, T.; Morioka, K.; Ogawa, M.; Yamagishi, S. Inducible oxacillin-hydrolyzing penicillinase in *aeromonas hydrophila* isolated from fish. *Antimicrob. Agents Chemother.* **1976**, *10*, 191–195. [CrossRef] [PubMed]

73. Bousse, L.; Mostarshed, S.; van der Schoot, B.; de Rooij, N.F. Comparison of the hysteresis of Ta_2O_5 and Si_3N_4 pH-sensing insulators. *Sens. Actuators B* **1994**, *17*, 157–164. [CrossRef]

74. Chou, J.C.; Wang, Y.F. Preparation and study on the drift and hysteresis properties of the tin oxide gate ISFET by the sol–gel method. *Sens. Actuators B* **2002**, *86*, 58–62. [CrossRef]

75. Product Information Urease, Type IX. Available online: https://www.sigmaaldrich.com/deepweb/assets/sigmaaldrich/product/documents/278/195/u4002dat.pdf (accessed on 24 November 2021).

76. Morais, P.V.; Gomes, V.F.; Silva, A.C.; Dantas, N.O.; Schöning, M.J.; Siqueira, J.R. Nanofilm of ZnO nanocrystals/carbon nanotubes as biocompatible layer for enzymatic biosensors in capacitive field-effect devices. *J. Mater. Sci.* **2017**, *52*, 12314–12325. [CrossRef]

77. Katz, E.; Minko, S. Enzyme-based logic systems interfaced with signal-responsive materials and electrodes. *Chem. Commun.* **2015**, *51*, 3493–3500. [CrossRef] [PubMed]

78. Katz, E.; Poghossian, A.; Schöning, M.J. Enzyme-based logic gates and circuits-analytical applications and interfacing with electronics. *Anal. Bioanal. Chem.* **2017**, *409*, 81–94. [CrossRef]

79. Wang, J.; Katz, E. Digital biosensors with built-in logic for biomedical applications—Biosensors based on a biocomputing concept. *Anal. Bioanal. Chem.* **2010**, *398*, 1591–1603. [CrossRef]

80. Lai, Y.H.; Sun, S.C.; Chuang, M.C. Biosensors with built-in biomolecular logic gates for practical applications. *Biosensors* **2014**, *4*, 273–300. [CrossRef]

81. Luo, C.; He, L.; Chen, F.; Fu, T.; Zhang, P.; Xiao, Z.; Liu, Y.; Tan, W. Stimulus-responsive nanomaterials containing logic gates for biomedical applications. *Cell Rep. Phys. Sci.* **2021**, *2*, 100350. [CrossRef]

82. Poghossian, A.; Malzahn, K.; Abouzar, M.H.; Mehndiratta, P.; Katz, E.; Schöning, M.J. Integration of biomolecular logic gates with field-effect transducers. *Electrochim. Acta* **2011**, *56*, 9661–9665. [CrossRef]

83. Poghossian, A.; Katz, E.; Schöning, M.J. Enzyme logic AND-Reset and OR-Reset gates based on a field-effect electronic transducer modified with multi-enzyme membrane. *Chem. Commun.* **2015**, *51*, 6564–6567. [CrossRef] [PubMed]

84. Honarvarfard, E.; Gamella, M.; Poghossian, A.; Schöning, M.J.; Katz, E. An enzyme-based reversible controlled NOT (CNOT) logic gate operating on a semiconductor transducer. *Appl. Mater. Today* **2017**, *9*, 266–270. [CrossRef]

85. Jablonski, M.; Poghossian, A.; Molinnus, D.; Keusgen, M.; Katz, E.; Schöning, M.J. Enzyme-based XOR logic gate with electronic transduction of the output signal. *Int. J. Unconv. Comput.* **2019**, *14*, 375–383.

86. Privman, V.; Zhou, J.; Halámek, J.; Katz, E. Realization and properties of biochemical-computing biocatalytic XOR gate based on signal change. *J. Phys. Chem. B* **2010**, *114*, 13601–13608. [CrossRef] [PubMed]

87. Schneider, A.; Eber, F.J.; Wenz, N.; Altintoprak, K.; Jeske, H.; Eiben, S.; Wege, C. Dynamic DNA-controlled "stop-and-go" assembly of well-defined protein domains on RNA-scaffolded TMV-like nanotubes. *Nanoscale* **2016**, *8*, 19853–19866. [CrossRef] [PubMed]

 *biosensors*

Review

Advances in the Detection of Dithiocarbamate Fungicides: Opportunities for Biosensors

Pablo Fanjul-Bolado [1], Ronen Fogel [2], Janice Limson [2], Cristina Purcarea [3] and Alina Vasilescu [4,*]

1 Metrohm DropSens, S.L., Vivero de Ciencias de la Salud, 33010 Oviedo, Spain; pablo.fanjul@metrohm.com
2 Biotechnology Innovation Centre, Rhodes University, Makhanda 6140, South Africa; r.fogel@ru.ac.za (R.F.); j.limson@ru.ac.za (J.L.)
3 Institute of Biology, 296 Splaiul Independentei, 060031 Bucharest, Romania; cristina.purcarea@ibiol.ro
4 International Centre of Biodynamics, 1B Intrarea Portocalelor, 060101 Bucharest, Romania
* Correspondence: avasilescu@biodyn.ro; Tel.: +40-21-310-4354

Abstract: Dithiocarbamate fungicides (DTFs) are widely used to control various fungal diseases in crops and ornamental plants. Maximum residual limits in the order of ppb-ppm are currently imposed by legislation to prevent toxicity problems associated with excessive use of DTFs. The specific analytical determination of DTFs is complicated by their low solubility in water and organic solvents. This review summarizes the current analytical procedures used for the analysis of DTF, including chromatography, spectroscopy, and sensor-based methods and discusses the challenges related to selectivity, sensitivity, and sample preparation. Biosensors based on enzymatic inhibition demonstrated potential as analytical tools for DTFs and warrant further research, considering novel enzymes from extremophilic sources. Meanwhile, Raman spectroscopy and various sensors appear very promising, provided the selectivity issues are solved.

Keywords: dithiocarbamate fungicides; chromatography; Raman spectroscopy; sensors; enzyme inhibition; voltammetry; biosensors

Citation: Fanjul-Bolado, P.; Fogel, R.; Limson, J.; Purcarea, C.; Vasilescu, A. Advances in the Detection of Dithiocarbamate Fungicides: Opportunities for Biosensors. *Biosensors* 2021, 11, 12. https://doi.org/10.3390/bios11010012

Received: 30 November 2020
Accepted: 27 December 2020
Published: 30 December 2020

Publisher's Note: MDPI stays neutral with regard to jurisdictional clai-ms in published maps and institutio-nal affiliations.

1. Introduction

Dithiocarbamate fungicides (DTFs) are non-systemic pesticides that have been used since the 1940s to control a number of fungal diseases in various crops and ornamental plants. Propineb, zineb, maneb, thiram, and mancozeb are amongst some of the most well-known and used fungicides, the chemical structures of which are presented in Figure 1.

While the development of new pesticide molecules and formulations has continued over the years, currently used DTFs such as maneb, mancozeb, propineb, thiram, and ziram were introduced more than 50 years ago (Figure 2) [1]. Mancozeb, propineb, and thiram are among the top selling fungicides, e.g., mancozeb sales are expected to reach $18 billion by 2025 [2].

Based on their chemical structure, DTFs are classified as propylene-bis-dithiocarbamates (PBs, e.g., propineb), ethylene-bis-dithiocarbamates (EBs, e.g., mancozeb, maneb, and zineb), and dimethyl dithiocarbamates (DDs, e.g., thiram, ziram, and ferbam). The DTFs from different groups have different toxicity, resulting in variances in the risk assessment for exposure to specific fungicides. Due to their toxicity, fungicides like zineb were banned in many countries around the globe, including the US and the EU, while in countries where it is currently allowed for use, maximum residue limits in the range of ppm are imposed by various organizations worldwide, for various food and agricultural products. The European Commission established maximum residue limits (MRL) of 0.01–25 ppm for dithiocarbamates in various plants and products of vegetable or animal origin (expressed as CS_2, including maneb, mancozeb, metiram, propineb, thiram, and ziram) [3]. Excessive use of DTFs has continued in recent years, e.g., amounts higher than the MRL have been detected in tomatoes [4], kiwi, and pears [5], etc.

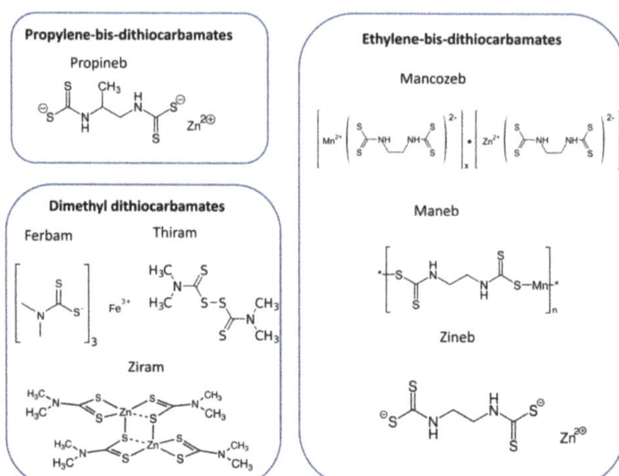

Figure 1. Classification of dithiocarbamate fungicides (DTF) and chemical structures of main representatives from each group.

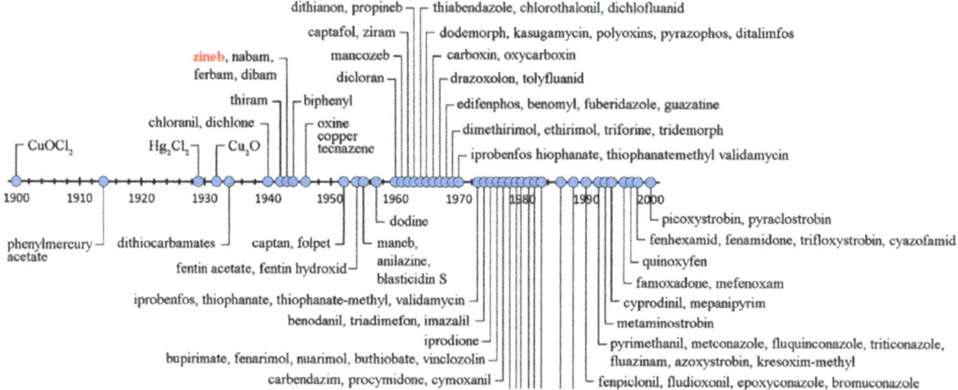

Figure 2. Timeline of the development of fungicides, showing the appearance of DTFs from 1930s onwards. Reproduced from [1], with permission.

Dithiocarbamate pesticides have low solubility in water and a number of organic solvents; hence, they are not typically extracted and analyzed by the multiresidue chromatographic methods used for screening other pesticides. This limitation, coupled to the wide number of DTF compounds applied as pesticides (Figure 2) meant that the simplest method for the analysis of dithiocarbamates relies on their degradation in acid media and analysis of the resulting CS_2 by spectrophotometric or chromatographic methods [3]. A major drawback to the detection of DTFs by quantification of formed CS_2 is the lack of specificity, as this does not allow the identification of parent DTFs present within the sample. Additionally, the analysis of DTFs based on CS_2 is affected by false positive results in agricultural products containing high levels of organic sulfur compounds: notably, CS_2 produced in acidic media by *Brassicaceae* and *Alliaceae* vegetables (e.g., cabbage and onion) was identified [3].

In a 2017 report, the European Food Safety Agency EFSA reiterated the need to develop specific analytical procedures for each active substance in this group of fungicides [5]. Currently, specific, single residue methods are available for thiram, propineb, and ziram [3].

Details of the various analytical procedures available for the determination of DTFs are given below, together with a discussion of the recent progress and perspectives in this field. A particular focus is placed on sensors and biosensors with the potential of delivering simple and fast detection.

2. Advances in DTFs Detection

2.1. Standard Chromatographic Methods for DTFs Detection

As with many other pesticide classes, the most selective methods for the determination of DTFs are based on gas chromatography coupled with-mass spectrometry (GC-MS) [6] and on reverse phase liquid chromatography coupled with optical, electrochemical, or mass-spectrometry detectors [7–10].

Nakamura et al. described a GC-MS method for the detection of 10 dithiocarbamates in foods [6]. The compounds were extracted as water-soluble sodium salts with cysteine and ethylenediaminetetraacetic acid (EDTA) and then were derivatized by methylation. This strategy enables differentiation between the DTFs from different groups, i.e., thiram, ziram, ferbam, zineb, maneb, mancozeb, milneb, metiram, propineb, nickel bis(dithiocarbamate), and polycarbamate. The quantification limits, expressed as CS_2 were 0.01 mg/kg in brown rice, soybean, potato, spinach, cabbage, apple, orange, pumpkin, cacao, cattle muscle, cattle fat, cattle liver, salmon, eel, milk, chicken egg, honey, and shrimp and 0.1 mg/kg in green tea samples [6].

An alternative approach based on reverse-phase liquid chromatography circumvented the need for derivatization [10], being based on the formation of ion pairs between the DTFs anions and tetrabutylammonium cations in alkaline media containing EDTA. The approach used two detectors connected in series, UV and electrochemical, achieving limits of quantitation, of 9, 12, 8 and 12 µg/L CS_2 for N-methyl-DTF, N,N-dimethyl-DTF, ethylenebis-DTF, and propylenebis-DTF [10]. However, while this method appears simpler and provided good recoveries from spiked fruit and tomatoes, accurate results can only be obtained for surface-intact vegetables.

Chromatographic methods developed with the aim of achieving sensitive and selective detection of DTFs relied not only on mass-spectrometry, UV, and electrochemical detectors but were also coupled with atomic absorption spectrometers. Indeed, the presence of different metals like zinc, manganese, and nickel in the structure of some DTFs makes it possible to identify the DTFs by atomic absorption spectrometry. Thus, procedures requiring high instrumental infrastructure like HPLC-UV with detection at 272 nm coupled with atomic absorption spectrometry were used for the determination of 10 pesticides, and were able to distinguish between zineb, maneb, and mancozeb in diverse matrices [11].

More recently, an LC-MS method was described for the analysis of 10 DTFs in beer, fruit juice, and malt samples, based on the common strategy of transforming the fungicides in water soluble salts and derivatizing them with methyl iodide [8]. The extraction of methylated derivatives of the DTFs was performed using a "quick, easy, cheap, effective, rugged, and safe" (QuEChERS) method, subsequently purifying the extracts by dispersive solid-phase extraction prior to LC-MS analysis. Separation of the methyl derivative DTFs compounds by Ultra Performance Liquid Chromatography (UPLC) used a C_{18} column and the detection was performed by ESI-MS in selected reaction monitoring (SRM), positive ion mode. The quantitation limits reached by this method for three representatives of the main groups of DTFs, i.e., propineb (a propylene-bis-dithiocarbamate), mancozeb (an ethylene-bis-dithiocarbamate), and thiram (dimethyl dithiocarbamate) were 0.52, 0.55 and 6.97 µg/kg, respectively. The procedure was successfully applied for the determination of DTFs in beer and fruit juice. A similar approach but with derivatization to dimethyl derivatives was described by Li et al., who reported detection limits of 0.6–1.6µg/kg and 0.8–2.5 µg/kg for mancozeb and propineb, respectively, in different vegetable food matrices [9].

While the selective detection of relevant DTFs at sensitivities below those required for the ppm MRL established for foodstuffs has been reported for several applicable matrices [7–10,12–14], the main limitations to the use of chromatographic techniques in routine detection of DTFs remain the cost of equipment and the relative complexity of analysis. Despite advances made in the field of portable GC-MS and HPLC chromatographic devices [15], the base costs of this equipment remain important (~US $100,000 or more). Similarly, the reliance of these techniques on purified organic solvents and the complex protocols required to satisfactorily detect DTFs with these approaches, including derivatization, purification, and pre-concentration of samples prior to the analysis (e.g., [8]) may prove prohibitive when attempting to apply these techniques at the scale required for the routine detection of DTFs in agricultural products. In summary, despite general progress in the development of cleaning cartridges, QuEChERS extraction methods and analytical instrumentation for chromatography and MS, there was not a huge advancement in the last years regarding the determination of DTFs.

2.2. Spectroscopy-Based Analysis Methods

Raman spectroscopy provides information about the vibrational states of molecules and therefore their functional groups and chemical structure. It is considered a fingerprint technique since each Raman spectrum corresponds to a unique chemical compound and a spectra library repository can provide rapid identification of molecules for analytical purposes. This highly selective technique is balanced by a lack of sensitivity due to the standard weak Raman scattering signals that usually provides detection limits at concentrations in the order of 10^{-2} M.

A great enhancement of the Raman signal can be obtained by the interaction of the analyte with some metallic surfaces mainly made of gold, copper or silver. Surface enhanced Raman scattering (SERS) was firstly introduced by Fleishman et al. in 1974 working with pyridine as a model probe and an electrochemically roughened silver electrode as surface [16]. The SERS effect increases the interest of Raman spectroscopy in analytical chemistry due to the sensitivity increase that can led towards to the detection of a single molecule [17].

First peak spectra assignments of dithiocarbamates (DMDTC and DEDTC) over silver surfaces (colloids and surfaces) were obtained in the 1990s by Mylrajan [18] and Tse Yuen [19], including FT-IR data.

For quantitative analysis several works have been done using as analytical signal the strongest band at around 1380 cm^{-1} assigned to the C-N stretching mode and symmetric CH$_3$ deformation mode of dithiocarbamates pesticides. Most of these approaches rely on silver or gold based nanomaterials to get the SERS enhancement:

Thiram was analyzed using silver nanoparticles clusters for SERS analysis, reaching a detection limit of 24 ppb (10^{-7} M) that is much lower than the maximum residue limit ranging from 2 to 15 ppm in fruit prescribed by the U.S. Environmental Protection Agency (EPA) [20]. A combination of silver nanocubes with reduced graphene oxide in a sponge-like structure was used for the detection of thiram and ferbam achieving detection limits of 10 and 16 ppb, respectively, based on the intensity of the characteristic signal at 1382 cm^{-1} [21]. In this case (Figure 3A) graphene oxide was used to remove the interference from aromatic pesticides adsorbed to it and allowing the SERS effect in the silver nanocubes to be detected only with DTF pesticides. As SERS spectra of ferbam and thiram were similar in terms of peak location and intensity ratios, principal component analysis (PCA) was used to distinguish which fungicide is present in the environment (see Figure 3B). It should be noted that the tolerances for ferbam residues in pear, apple, grape, mango, cabbage, and lettuce, range from 4 to 7 ppm, much higher concentrations than the ppb level achieved with this approach.

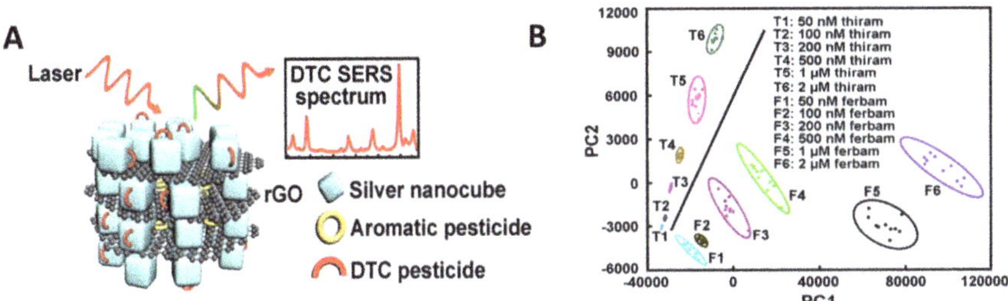

Figure 3. (**A**) Principle of the surface enhanced Raman scattering (SERS) detection of DTF using rGO-wrapped Ag nanocubes. (**B**) Principal component analysis (PCA) plot (PC1 versus PC2) of signals extracted from 10 SERS spectra of thiram and ferbam with concentration ranging from 50 nM to 2 µM. Reproduced from [21] with permission.

The results revealed that while there is a clear differentiation between the two DTFs based on PCA when analyzing individual spectra, it was not possible to distinguish between them when present in mixtures. Still, the total amount of the S−C−S group in the fungicide mixture was correlated with $I^{0.5}$, where I is the intensity of the SERS signal at 1382 cm^{-1}. Other recently published papers related to the detection DTFs usually work with thiram as probe analyte using a variety of silver nanomaterials for the SERS effect [22–24].

Gold nanorods were used for ultra-sensitive detection of DTFs [25,26]. Thiram, ferbam, and ziram were determined in the range of low ppb concentrations (10^8 M). The interaction of the fungicide with the gold substrate is supposed to undergo spontaneous cleavage of their metal–sulfur bonds to produce the dimethyldithiocarbamate ion which then assembles on the substrate surface. Since the degradation processes produce identical ions for all these molecules, the SERS spectra of the three pesticides appear very similar; therefore, multivariate data analysis techniques can be coupled when working with real samples [25]. For additional information of the spontaneous assembly of dithiocarbamate ligands on gold metal substrates, a research paper by Alexander Wei et al. [27] can be revisited. Gold nanoparticles trapped into cellulose matrices has been reported as an interesting solution for the in situ extraction and detection of thiram in residues in soil and fruits [28]. In addition, screen printed gold electrodes roughened through an electrochemical pretreatment arise as an easy to use and portable solution for the detection of thiram and other pesticides as well [29].

Efficient sampling, enabling fast and quantitative recovery of DTFs is a limiting factor for the practical applications of SERS. In this respect significant progress was reported in the last years with regards to flexible substrates that can be brought in close contact with sample surface, with double role of collecting the DTFs from samples and as SERS substrate. This includes an approach for the in situ extraction and SERS substrate formation [28], applied for samples with irregular surfaces such as soil, strawberries, and cucumbers as well as swab-type devices, e.g., applied for sampling thiram from the surface of intact fruits and vegetables [24].

The detection of thiram was demonstrated in a variety of spiked samples, including soil, strawberries, tomato, cucumber, water, etc., for which satisfactory recoveries were calculated.

Raman spectroscopy is one of the best suited techniques for the determination of dithiocarbamates due to the high selectivity provided by the fingerprint spectrum of each molecule and the high sensitivity achieved through the SERS effect. Additionally the potential use of portable analyzers and cost effective disposable SERS substrates allows its application in field analysis. However, some different sensing alternatives based on other optical detection modes are reported showing advantages and drawbacks for dithiocarbamate determination.

Fourier transform infrared (FTIR) spectrometric procedures were developed for the determination of ziram and thiram in solid samples [30] and in vapor phase samples, after ziram decomposition [31]. Both FTIR-based methodologies allowed for pesticide detection in the order of tens and hundreds of mg as absolute detection limits.

From our perspective the rapid evolution of Raman spectrometers from benchtop instruments to portable devices and the cost reduction of the technology makes it very powerful for "in situ" analysis. Moreover the huge number of cost-effective SERS substrates developed using different nanomaterials position this technique as the leader in terms of sensitivity. The intrinsic selectivity provided by the Raman spectrum avoids the need of any other recognition element. However, when several DTFs are present in the same sample, chemometrics methods have to be applied to SERS data in order to specifically detect individual fungicides. Sample collection and sample immobilization in the SERS substrate to get quantitative analysis are still the main drawbacks to be solved.

2.3. Optical and Electrochemical Assays

Electrochemical sensors and optical assays have been investigated as simpler, miniaturized, and cost-effective analytical devices that could represent viable alternatives to the chromatographic methods and Raman spectroscopy.

2.3.1. Electrochemical Sensors

The electrochemical detection of dithiocarbamate fungicides has been well-known for several decades [32,33], facilitated by the multiple electroactive sites present on these compounds (Figure 4A).

Under aqueous conditions, detection via the thiol groups is facilitated by the dissociation of metal-complexed dithiocarbamates to produce carbamate anions (Figure 4A) [34]. At mercury electrodes, carbamate anions are readily reduced to form mercury complexes (e.g., Figure 4B). At inert electrodes (such as carbon and platinum), carbamate anions are broadly irreversibly oxidized (e.g., Figure 4C) by a monoelectron step [33]. This forms radical intermediates that frequently dimerize via the sulfur atoms to form disulfide products [33]; this complex being further oxidized at higher anodic potential [33,35].

Apart from the particular structure of dithiocarbamate under investigation, the exact detected moiety/ies can be controlled by the selection of experimental parameters, such as electrode type, electrolyte solvent selection (especially pH) and the electrochemical waveform applied to their detection. This has facilitated the detection of numerous commercially-used dithiocarbamates in multiple studies (Table 1), in both laboratory-formulated pure standards and within real samples (predominantly commercial formulations of the pesticides, and pesticide-spiked foodstuffs).

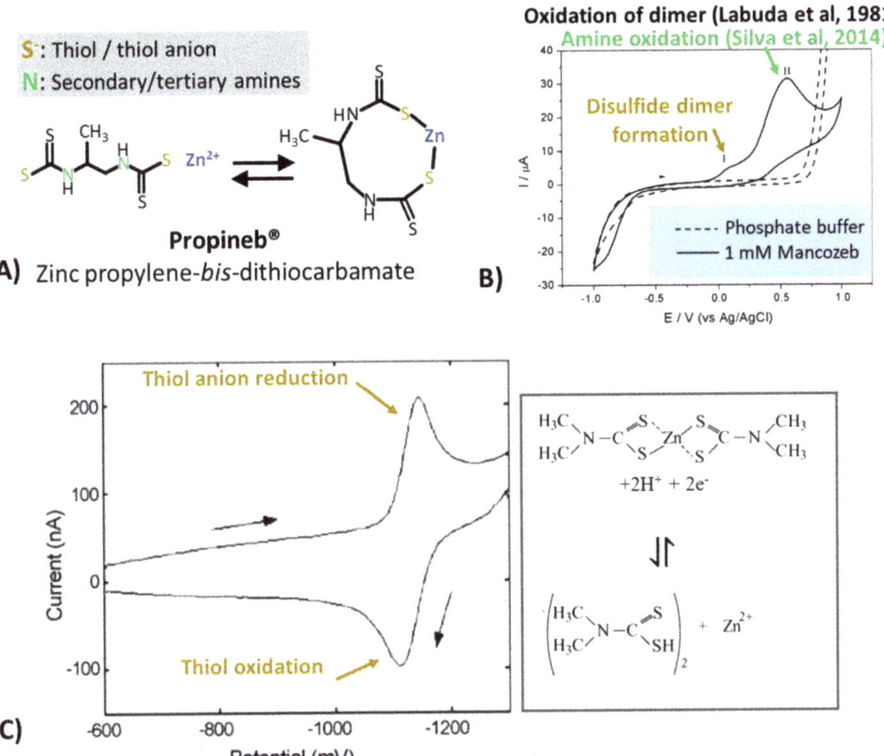

Figure 4. Electrochemical detection of dithiocarbamate pesticides is facilitated by the multiple electrochemically-active moieties they possess. (**A**) Commonly-reported electroactive moieties of dithiocarbamates, using Propineb® (extracted from [36]) as a general example of this class of compound. (**B**) Cyclic voltammograms of Mancozeb at a Boron-doped diamond electrode in phosphate buffer (extracted from [36]), showing anodic processes involving its thiol and/or amine moieties: two separate possible mechanisms behind the more positive peak are shown, one cited in [36] and that discussed by [35]. (**C**) Cyclic voltammogram of Ziram at a Hanging Mercury Drop Electrode in Britton–Robinson buffer, pH 2.8 (extracted from [37]. Peaks are attributed to the reversible reduction of the thiol moieties and subsequent displacement of the zinc moiety (inset, from same reference) at mercury electrodes. Reproduced from [36,37], by permission).

Table 1. Direct electrochemical detection: selectivity via catalysis and/or electrode potential control.

Dithiocarbamate Pesticides Investigated	Electrode Surface (Catalyst/Modifier) [a]	Real Samples Investigated	Signal Basis [b]	L.O.D. (Analytical Ranges Reported)	Ref.
Ziram	Polished silver solid amalgam electrode	Spiked river waters	SWV	0.24 μM	[38]
Thiram	Hg		CS-DPV peak at −0.55 V vs. Ag/AgCl	0.12 μM	[39]
Thiram	Rotating gold disk electrode	Commercial formulations; spiked water samples	Ads-LSV, peak at +1.4 to +1.5 V vs. Ag/AgCl	16 nM	[40]
Thiram Disulfiram	Graphite-PTFE composite electrode	Extracts of spiked strawberry samples	Ads-LSV, peaks at +0.85 V vs. SCE	Thiram: 54 nM (0.2 to 1 μM) Disulfiram: 20 nM (0.2 to 1 μM)	[41]
Thiram Disulfiram	Graphite-PTFE composite electrode	Spiked tap and well water samples	FIA-CA at +1V vs. Ag/AgCl	Thiram: 43 nM (0.1 to 1 μM) Disulfiram: 20 nM (0.1 to 1 μM)	[42]
Ziram	Hg	Extracts of spiked rice samples	CS-DPV	32 nM i.e., 10 ppb	[43]
Zineb	Hg		AdSV, cathodic peak at −0.455 V vs. Ag/AgCl	1 nM	[44]
Carbathion, Ferbam, Nabam, Thiram, Thiuram, Zineb, Ziram	Carbon paste electrode - Fe(II)metallophthalocyanine composite		Ads-LSV	Ranged from 10 nM (carbathion) to 200 nM (Thiuram)	[45]
Nabam	GCE, modified with Co(II) phthalocyanine and carbon ink		LSV, peak at −0.2V vs. Ag/AgCL	28.8 nM	[46]
Thiram	GCE	Commercial formulations; plant sample extracts exposed to thiram	SWV at +0.34 V vs. Ag/AgCl	n.r.	[47]
Carbathion	GCE		CV, peak forming at +1.46 vs. Ag/AgCl	9.3 μM (132 μM to 224 μM)	[48]
			SWV, peak forming at +1.46 vs. Ag/AgCl	85 nM (2 μM to 7.7 μM))	
			FIA-CA potential of +1.3 V vs. Ag/AgCl	10 nM (1.2 μM to 6 μM)	
Ziram	Hg	Extracts of spiked vegetable samples	SWV, −1.1V vs. Ag/AgCl.	23 nM (33 to 328 nM)	[37]
Thiram	Copper-mercury amalgam electrode	Spiked river water samples	CS-SPV, peak between −0.59 and −0.8 V vs. Ag/AgCl	16 nM	[49]
Propineb	Carbon-paste electrode (Cu²⁺-enriched montmorillonite)	Commercial formulation	Ads-SWV, peak at ~−0.1V vs. SCE	1 μM	[50]
Mancozeb	BDD		PAD at +0.3V vs. Ag/AgCl)	0.514 μM (40 to 650 μM)	[36]
Mancozeb	GCE	Commercial formulation	Ads-SWV, peaks forming at −0.7V vs. Ag/AgCl	7 μM	[51]
Ziram	BDD	Spiked river water samples	FIA-CA at +0.55 V	2.7 nM	[52]

Table 1. *Cont.*

Dithiocarbamate Pesticides Investigated	Electrode Surface (Catalyst/Modifier) [a]	Real Samples Investigated	Signal Basis [b]	L.O.D. (Analytical Ranges Reported)	Ref.
Maneb	BDD	River water	DPV peak at +0.9V vs. Ag/AgCl	24 nM (80 nM to 3 μM)	[53]
Mancozeb	Single-crystal (Au(111) and Au(110)		Ads-LSV, peaks at −0.6 to −0.96V vs. Ag/AgCl	Au(110): 100 nM Au(111): 500 nM	[54]
Mancozeb	Gold electrode modified with Poly (3,4-ethylene dioxythiophene), multi-walled carbon nanotubes, and gold nanoparticles	Water	CV, anodic peak +0.65 V vs. Ag/AgCl	5 μM	[55]
Thiram	Carbon paste electrode modified with zeolite	Aqueous extracts of fruit juices	DPV, anodic wave at +0.70V vs. Ag/AgCl;	4 nM (14 nM to 4.2 μM)	[56]
Thiram	Platinum, modified with silver nanoparticles	Tap, canal, and river water	DPV and CV	0.731 μM or 0.18 ppm	[57]
Thiram	GCE (dissolved Zn^{2+} and Cu^{2+} cations)	River water	CS-LSV: −1.330 V vs. Ag/AgCl for Zn-Thiram; +0.020V for Cu-Thiram complexes.	n.r. (5 to 50 μM)	[58]

[1] Where possible, analytical parameters have been standardized: limits of detection and analytical ranges of reported sensitivities are standardized to mol/L (M) units and detection sensitivities of amperometric and voltametric signals (i.e., peak currents and or response currents) were standardized to μA/μM values; [a]—Electrode surfaces:BDD: Boron-doped diamond; GCE: Glassy carbon electrodes; Hg: Mercury (Drop) Electrodes; PTFE: poly(tetrafluoroethylene); [b]—Signal basis:Ads-: Adsorptive (prefix); AS—Anodic Stripping (prefix); CS—Cathodic Stripping (prefix); FIA—Flow-injection analysis (prefix);CA: Chronoamperometry CV: Cyclic Voltammetry; DPV—differential pulse voltammetry; LSV: Linear Sweep Voltammetry; PAD: Pulsed Amperometric Detection; SCE: Saturated calomel electrode; SWV: Square-wave voltammetry; IUPAC designations of commercial dithiocarbamates studied in published articles: Carbathion: sodium N-methyldithiocarbamate (also known as metam sodium); Disulfiram: tetraethylthiuram disulfide; Diram: sodium N,N-dimethyldithiocarbamate; Ferbam: iron N,N′-dimethyldithiocarbamate; Maneb: Manganese ethylene-*bis*-dithiocarbamate; Mancozeb: 1,2-ethanedicarbamic acid, tetrathio- Manganese Zinc ethylene-*bis*-dithiocarbamate); Metam sodium: sodium N-methyldithiocarbamate; Nabam: sodium N,N′-ethylene-*bis*-dithiocarbamate; Propineb: Zinc propylene 1,2-*bis*-dithiocarbamate; Thiram: tetramethylthiuram disulfide; Thiuram: tetraethylthiuram disulfide; Zineb: zinc N,N′-ethylene-*bis*-dithiocarbamate; Ziram: N,N′-dimethyldithiocarbamate.

As is evident by the large number of recent publications, the application of electrochemical methods of detection of dithiocarbamates remains an actively-researched field of enquiry. This is due to a combination of their aforementioned ready detection via electrochemical means and the inherent sensitivity of electrochemical detection methods. A number of publications report the sensitive detection of different dithiocarbamate fungicides: analytical quantification ranges of reports are frequently in the μM-to-nM orders of magnitude and reported limits of detection often in the nM range (Table 1).

The large variation in the dithiocarbamates studies, the working electrodes and waveforms in the literature reviewed makes direct comparison across all the entries difficult, but nonetheless, some trends are evident across the literature.

At high concentrations of unionized forms of DTC pesticides, the low aqueous solubility remained a continual analytical challenge. Most concentrated stocks were formulated in the μg/mL range, typically achieving this by including chelating agent, most often EDTA, e.g., [36,51] and high pH in aqueous solvents to improve solubility; sometimes with combinations of the two [59]. Others make use of organic solvents: acetonitrile (e.g., [56], chloroform [59], methanol [12,39,40,42], and ethanol [47] to solubilize these.)

The low aqueous solubility of these pesticides somewhat improves sensitivity by electrochemical methods, due to the tendency of target molecules to adsorb to the surface of electrodes, effectively preconcentrating the analyte before quantification. To capitalize on this phenomenon, most reported investigations explicitly make use of adsorptive waveforms (Table 1): dithiocarbamates have been routinely detected using adsorptive stripping voltammetry waveforms, e.g., [51], cathodic stripping voltammetry [43], anodic stripping voltammetry [45] or other forms that deliberately include an adsorptive step prior to analysis (e.g., pulsed amperometric detection by [36]. The majority of electrode surfaces selected tend to be those extensively used for adsorptive studies: aromatic and graphitized carbon preconcentrating the pesticides using the hydrophobic nature of the ethylene backbones. A more specific targeting of the thiol moieties of thiocarbamates has been reported, either using gold electrodes (e.g., [54] to form thiol-gold self-assembled adlayers at the electrode surface or the formation of mercury amalgams at mercury drop-electrodes [34].

As previously discussed, mercury drop electrodes were frequently employed for the detection of dithiocarbamates. All of the identified studies made use of cathodic stripping voltammetry to detect and quantify their investigated dithiocarbamates, preceded by an adsorptive preconcentration. Several of these reports explicitly cited the reduction of thiol moieties as the basis for signal generation at these electrodes [37,44,49]. A similar catalytic effect using copper–mercury amalgam electrodes was reported for monitoring of thiram [49]. Despite the overall sensitivity of this approach (the majority of studies reporting limits of detection in the low nM range), the use of mercury electrodes in routine electrochemistry has declined in recent years, due to the associated health and environmental hazards of this metal and may preclude its use in the commercial detection of these fungicides.

The detection of dithiocarbamates by unmodified carbon electrode surfaces (i.e., entries reported to use boron-doped diamond, glassy carbon electrodes, carbon-paste electrodes) form the largest group found during this review. Similar to expectations, these predominantly use anodic processes to detect their targets, for reasons discussed previously. While inherently less sensitive than the cathodic approach for mercury drop electrodes (stated limits of detection using voltametric approaches with carbon surfaces typically reported at µM levels in Table 1), optimization of the detection waveform used and other experimental conditions can enhance sensitivity to the nM level (e.g., [48,59] in Table 1), similar to those achieved using mercury drop electrodes.

Many of the detection methods investigated using unmodified metal and carbon electrodes share a disadvantage: the application of substantial cathodic or anodic voltages required to effectively detect dithiocarbamates. These substantially decrease the signal specificity used, as numerous other compounds capable of electrochemical reactions at more negative working electrode potentials (for anodic detection) or more positive (for cathodic detection) will also contribute signal and decrease selectivity. Several of these reports [59] reported that operating at electrode potentials where Faradaic processes in the electrolyte occurred began affecting signal-to-noise ratios due to this limitation. Despite this, numerous reports in Table 1 also include samples extracted from foodstuffs, soil and river water samples and report satisfactory detected recoveries from these using the above analytical approaches.

To further improve selectivity, reverse phase liquid chromatographic separation of sample components has been reported as a means of coupling the sensitivity of amperometric detection using unmodified electrodes with some means of separating out the various components of the samples (Table 2). This allowed for several investigations to separately measure dithiocarbamates in a mixture of these pesticides.

Table 2. Chromatography-coupled detection: selectivity via chromatographic separation of sample components.

DTF Investigated	Electrode Surface (Catalyst/Modifier)	Sample	Applied Potential	L.O.D.	Ref.
Thiram	CPE	Spiked river water	+1.1V vs. Ag/AgCl	2.07 µM,	[12]
Thiram Disulfiram	Composite PTFE-graphite paste electrodes	Spiked apple samples	+1V vs. Ag/AgCl	Thiram: 1.66 µM Disulfiram: 3.37 µM	[60]
Carbathion Thiram Zineb	GCE	Spiked fruit pulp samples	+1.1 V vs. Pd	0.7 µM Thiram: 1.5 µM Carbathion: 0.7 µM	[59]
Carbathion Mancozeb Propineb Ziram	not reported		+0.6V vs. Pd.	Carbathion: 31 nM Mancozeb: 7 nM Propineb: 26 nM Ziram: 26 nM	[10]
Thiram	GCE	Spiked tap water and beetroot juice	+1.4V vs. Ag/AgCl.	13.4 nM	[13]
Thiram, disulfiram	AuNP-SPCE	Spiked apple, grape and lettuce samples	+1.2 V vs. Ag/AgCl	Thiram: 91 nM Disulfiram: 0.56 µM	[14]

CPE: carbon paste electrode. GCE: glassy carbon electrode. AuNP-SPCE: Gold nanoparticle modified screen-printed carbon. PTFE: poly(tetrafluoroethylene).

Increasingly, possibly as a sustainable alternative to mercury electrodes, the more recent trend is the investigation of catalysts and/or electrode modifiers to improve signal specificity or sensitivity [50] of the assays. Some of these forms of modification attempt to enhance the adsorption of dithiocarbamates to enhance detection sensitivity e.g., PTFE [41,42] and mercury amalgams [49]. Others apply catalysts to facilitate electron-transfer between DTFs and the electrode: the most commonly-identified are metallophthalocyanines [45,46] and metal nanoparticles [55,57] to catalyze anodic detection of DTFs. While all of the catalysts reportedly impart enhanced sensitivity, none of them increase specificity: the broad electrocatalytic effects of metallophthalocyanines [61] and metal nanoparticles [62] are widely-known.

While overall good sensitivity is reported using electrochemical detection of DTFs, neither the electrode surfaces nor the waveforms selected are specific enough for routine detection and are inherently prone to multiple interferences. Despite the recent research focus on alternative catalysts and modifiers, a specific electrochemical DTF chemosensor is not yet apparent in the literature.

2.3.2. Optical Assays

Some simple colorimetric assays working with metallic nanoparticles based on copper [63], silver [64], or gold [65] can be used for the detection of pesticides due to the solution colour change after nanoparticles aggregation in presence of the analyte. In these approaches nanoparticle protective agents such as sodium dodecyl sulfate (SDS), cetyl tributylammonium bromide (CTAB) or citrate play a key role in the procedure that finally allows a semiquantitative detection by naked eye or a quantitative detection by using a simple spectrophotometer. Detection limits in the ppb level can be achieved in samples of environmental interest [65]. Ease of use and no need for expensive analytical instrumentation are the main advantages of this approach; however, a lack of selectivity can be highlighted as the main drawback. A multicolor sensor for visual screening of total dithiocarbamates based on the inhibition of gold nanoparticles growth in presence of the fungicides [66] clearly demonstrates these conclusions. Sufficient sensitivity, and short analysis time makes it useful for screening purposes but the lack of selectivity between ziram, thiram, and zineb makes it only useful for total DTCs detection.

Ziram was quantified with a detection limit of 2nM in tomatoes and rice using the same gold nanoparticle aggregation principle. In this case, the detection was made using a fluorimeter since a yellow fluorescence decrease of quantum dots was monitored because of a quenching mechanism [67]. Working with phosphorescent Mn doped ZnS quantum dots a quenching effect was also used for the detection of thiram at a detection limit of 50 nM. This chemosensor was tested in fruit peels and minor interferences were found with atrazine [68].

There are some other works related to the detection of thiram by electrochemilumi-nescence (ECL), but these cannot be considered in the range of sensor devices. Most of them are online analytical methodologies based on Flow Injection Analysis systems where the ECL analytical signal is enhanced in the presence of thiram [69,70]. Working with the same highly sensitive analytical technique (ECL), an ELISA assay was developed for the detection of thiram in honeybees. An indirect competitive assay was implemented and a detection limit of 9 ppb was achieved [71]. According to the state of art, there is still enough scope for the development of optical sensors working in the detection of dithiocarbamates fungicides, mainly combining specific recognition elements for these analytes with an optical transducer. In this case enzymatic, antibody, or aptamer based optical biosensors should be further developed.

2.4. Biosensors Based on Enzyme Inhibition

Coupling a specific biorecognition element with a sensitive detection method as is achieved in biosensors goes a long way towards improving the selectivity of detection and eliminating the need for chromatographic separation or complicated chemometric analysis. In addition, compared to electrochemical sensors for DTFs discussed in Section 2.3.1, operating at large overpotentials where interferences in real samples are important, electrochemical biosensors rely on electron transfer mediators or on direct electron transfer from the enzymes to the electrode. Electrochemical biosensors are thus operated at lower overpotentials, alleviating the problems related to nonspecific Faradaic reactions, common for electrochemical sensors.

However, the detection of DTFs was very little explored in the biosensing field compared to other classes of pesticides. While the first ELISA tests for the detection of thiram in a food matrix, i.e., in spiked lettuce was reported 20 years ago [72], no reports of biosensors based on antibodies followed. Moreover, there are no recent reports on specific antibodies and related kits for DTFs, no aptamers have been selected for specific DTFs and no sensors or sample extraction methods based on molecularly imprinted polymers have been reported for this type of pesticides. The only attempts to specifically detect DTFs with biosensors exploited enzyme inhibition and the toxic effects on mammalian cells as detailed in Section 2.4.1 below.

The detection of pesticides based on the principle of enzyme inhibition has long been explored as an alternative to chromatographic and spectroscopic methods. Various enzymes, including acetylcholinesterase, butyrylcholinesterase, alkaline phosphatase, organophosphorus hydrolase, the enzymatic complex of Photosytem II in plants, algae, and cyanobacteria, as well as urease, laccase, tyrosinase, and aldehyde dehydrogenase have been used for the detection of various classes of pesticides [73–76]. DTFs are known inhibitors of aldehyde dehydrogenase, tyrosinase, and laccase, which led to the development of several biosensors that make use of these enzymes [77–83].

In inhibition-based biosensors, the analytical signal is measured before and after exposure of the sensor to a sample containing the pesticide target; corresponding changes in signal are correlated to the concentration of the pesticide (Figure 5).

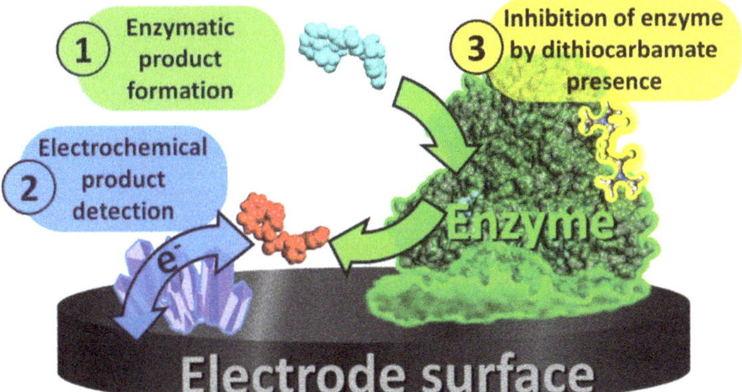

Figure 5. Generalized schematic of the operating principles of DTF-monitoring electrochemical biosensors. 1: During operation, enzymes immobilized at, or near an electrode, are supplied with their required cofactors and substrates which produce a constant rate of enzymatic product. 2: The electrochemical detection of the formed enzymatic products generates the biosensor's signal. 3: The presence of DTF pesticides inhibits the enzyme activity, lowering the electrochemical signal to an extent proportional to DTF concentration.

Depending on the precise enzyme and target under study, enzymatic inhibition is either reversible or irreversible and can follow either competitive, uncompetitive, non-competitive, or mixed types of inhibition mechanisms. Understanding the kinetics of the enzymatic reaction and the inhibition mechanism is very useful for developing an efficient biosensor [76]. Inhibition-based biosensors rely mainly on optical and electrochemical detection methods, similar to those discussed in Sections 2.3.1 and 2.3.2 above. A more detailed discussion can be found in [76] and [74], amongst others.

It is important to note that the enzymes employed are not inhibited by a specific compound, but by a range of pesticides belonging to specific chemical groups, as well as by other compounds such as heavy metals for example. This lack of selectivity has prompted some skepticism regarding the analytical opportunities of inhibition-based enzymatic biosensors [84]. Nonetheless, the potential advantages of such analytical devices (particularly, portability, and fast sensor responses) seem to prevail, making them suitable as rapid screening and alert systems. Consequently, these biosensors continue to attract interest. To address the lack of selectivity, multiplexed sensors coupled with chemometric analysis, genetically modified enzymes, as well as novel enzymes, e.g., extracted from extremophilic microorganisms, with different substrate specificity and inhibitor profile are continuously explored.

2.4.1. Examples of Biosensors for the Determination of DTFs

For the detection of DTFs, several electrochemical biosensors have been developed as presented in Table 3 and detailed further below.

Table 3. Examples of biosensors for DTF based on enzymatic inhibition.

Fungicide	Detection Method	Enzyme	Limit of Detection	Incubation Time	Reference
Ziram	Square wave voltammetry/GPE	LACC [1], adsorption on electrodeposited Prussian Blue film	0.002 ppm	15 min	[77]
Ziram	Square wave voltammetry/GPE	LACC-TYR-AuNPs -CS electrodeposited film	1 ppb	20 min	[79]
Maneb	Amperometry/Pt electrode	ALDH+DP, entrapment in PVA/SbQ	1.48 ppb	15 min	[81]
Zineb	Amperometry/Pt-sputtered SPCE	ALDH and NADH oxidase/entrapment in PVA/SbQ	8 ppm 8–80 ppb	5 min	[82]
MITC	Amperometry/MBRS SPCE	ALDH/entrapment in PVA/SbQ	100 ppm	10 min	[78]
Maneb and zineb	Chronoamperometry/MBRS-SPCE	ALDH/ entrapment in PVA/SbQ or cross-linking with glutaraldehyde	31.5 ppb–maneb 35 ppb-zineb	10 min	[80]
Propineb (and organophosphates)	Potentiometry/Ag coated with AgCl	Working electrode inserted into Calcium-alginate beads containing 5×10^4 cultured N2a or Vero mammalian cells.	0.33 µM (Vero cells) to 1.65 µM (N2a)	2.5 min	[83]

[1] Abbreviations: PVA-SbQ: poly(vinyl alcohol), bearing styrylpyridinium groups. MITC: Methyl Isothiocyanate. MBRS: Meldola Blue-Reinecke salt. GPE: Graphene doped carbon paste electrode. AuNO: gold nanoparticles. CS: chitosan SPCE: screen-printed carbon electrode. LACC: laccase. Tyr: tyrosinase. ALDH: aldehyde dehydrogenase. DP: diaphorase.

In a departure from most biosensor configurations, a whole-cell potentiometric sensor was developed using mammalian N2a (murine neuroblastoma) and Vero (green monkey kidney epithelial) cells entrapped in alginate beads, attached to a silver working electrode [73]. Detection of propineb in this format using the N2a cells was attributed to inhibition of membrane-bound acetylcholinesterases expressed by the mammalian cells and subsequent influence on the membrane polarization of these cells; a similar response was separately attributed to binding of the zinc-centered zineb to a zinc ion channel protein expressed by the Vero cells.

Laccases (polyphenol oxidase, EC 1.10.3.2) catalyze the oxidation of a wide range of aromatic organic compounds, including diphenols in the presence of molecular oxygen. The activity of laccase from *Trametes versicolor* is inhibited by ziram, but also by carbamate insecticides such as methomyl, pirimicarb, formetanate, carbaryl, and carbofuran [77,85,86] and by arsenate and arsenite. [87], among others. In a biosensor for the detection of ziram, laccase was dropcasted on a graphene doped carbon paste electrode, coated with a film of Prussian Blue [77]. When the biosensor was inserted in a solution of 4-aminophenol, upon sweeping the potential from +0.3 V to −0.1 V, 4-aminophenol was electrochemically reduced to an imine-quinone intermediate, that was further transformed into p-quinone in a reaction catalyzed by laccase. The cathodic peak current at −0.05 V due to the reduction of p-quinone, formed in the enzymatic reaction was proportional with laccase activity. Based on this principle, the biosensor enabled the analysis of ziram with a detection limit of 0.002 ppm, in a linear range between 2.49×10^{-8} M and 5.66×10^{-7} M. Several carbamate pesticides were also determined with detection limits of 0.001 ppm (carbaryl), 0.007 ppm (pirimicarb), 0.013 ppm (formetanate), and 0.022 ppm (carbofuran).

Tomato and potato extracts, spiked with pesticides at two concentration levels, were analyzed with the biosensor. Remarkably, a Quick, Easy, Cheap, Effective, Rugged, and Safe—QuEChERS method was used for pesticides extraction from the vegetable samples. The recoveries for ziram were 97.6–101.1% while for the other pesticides were 90.2–100.3%, indicating a satisfactory accuracy of the biosensor. In addition, the biosensor had a good reproducibility (RSD = 5.0%, n = 4 sensors) and repeatability (RSD = 1.8%, 2.4% for intra-day and inter-day repeatability) and was stable for 20 days. The selectivity study has focused on vitamins contained by tomatoes and potatoes (beta-carotene, i.e., pro-vitamin A, thiamine, i.e., vitamin B1 and ascorbic acid, i.e., vitamin C). Significant interferences, i.e., amounting to 11.4–13.2% of the analytical signal were found for beta-carotene and ascorbic acid, when tested at very high concentrations that were 10 times higher than those of spiked pesticides.

This report on a laccase-based biosensor provides a good illustration of both the potentialities and the challenges for the electrochemical biosensors for pesticides that rely on enzyme inhibition. The good sensitivity of the biosensor was attributed in part to the characteristics of the electrochemical transducer, obtained from carbon paste containing 20% graphene flakes and modified with an electrodeposited film of Prussian Blue. The graphene flakes had a thickness of few layers, a length of 500 nm–1.5 μm and contained 12.4% oxygen, 87.0% carbon, and 0.5% nitrogen. Modification of the carbon paste with this type of graphene improved the speed of the charge transfer in the electrochemical reduction of 4-aminophenol to the imine-quinone intermediate. Further modification of the electrode with Prussian Blue enhanced the cathodic current by approximately 27%. The use of nanomaterials to improve the conductivity and enhance the active surface area of the base transducer in biosensors is common nowadays, with more nanomaterials explored each day. Nonetheless, the characteristics of these nanomaterials vary in large limits, depending on the method used for obtaining them which dictates properties such as the degree of oxidation, number of stacked layers, size and possible contaminants, all influencing the electrochemical characteristics of the modified electrodes. While in this 2013 report [77] laccase was simply adsorbed on the electrode, controlled immobilization of laccase was also demonstrated in inhibition based biosensors for arsenate and arsenite [87]. Anthracene moieties covalently bound to multi-walled carbon nanotubes anchor laccase in a controlled manner, with the copper center of laccase oriented towards the electrode surface, thus allowing for direct electron transfer from enzyme to the electrode for the catalytic oxygen reduction. This immobilization approach also minimizes potential interferences due to chloride [87]. It is therefore reasonable to expect that better, more sensitive and selective inhibition-based biosensors can be obtained by exploiting the knowledge accumulated so far with regards of different nanomaterials and their coupling with enzymes and other modifiers. With regards to practical applications, the report of Oliveira et al., [77] emphasized the necessity of application-targeted investigations of possible interferences. Furthermore, QuEChERS methods have all the advantages for sample extraction denoted by their name but require lab-dedicated equipment. Sample pre-treatment remains therefore a bottleneck for the development of applications for in-field screening of DTFs and other pesticides. Moreover, definitive proof of biosensor accuracy should be obtained by comparing the results obtained with the biosensor with a standard confirmatory method or by using certified materials.

In addition to laccase, tyrosinase is another enzyme that is inhibited by DTFs. Tyrosinase (EC 1.14.18.1, monophenol, o-diphenol: oxygen oxidoreductase) catalyzes the oxidation of monophenols to o-diphenols, as well as the further oxidation of diphenols to their corresponding quinones. Tyrosinase is also inhibited by atrazine, hydrazines, and cyanide. In electrochemical biosensor investigations, tyrosinase was coupled with laccase in a biosensor for the detection of carbamate pesticides, achieving higher sensitivity than when each enzyme was used alone [79]. The mono and bi-enzymatic laccase biosensors were applied for the determination of ziram, as well as the carbamate pesticides carbofuran, formetanate carbaryl, and propoxur in spiked vegetables (tomato, potato) and citrus fruit (lemon, tangerine, and orange). According to the principle depicted in Figure 6A, 4 amino-phenol was used as an enzymatic substrate that was converted into p-benzoquinone under the catalytic action of tyrosinase and laccase. The magnitude of the cathodic current due to the reduction of p-benzoquinone to p-hydroquinone was correlated with enzyme activity. The cathodic current decreased in the presence of ziram in a concentration dependent manner as depicted in Figure 6B.

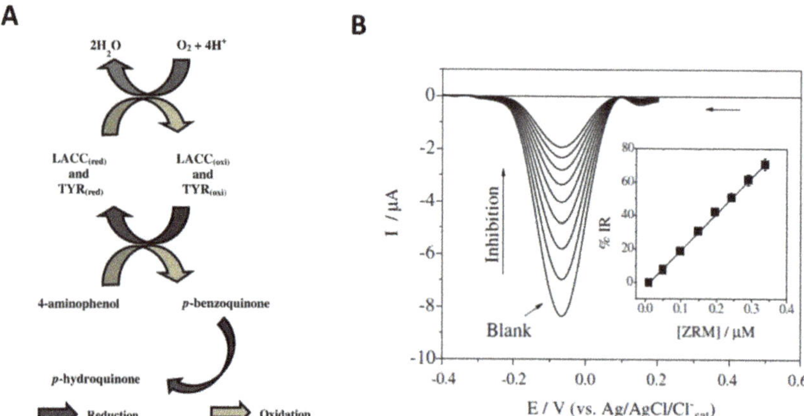

Figure 6. (**A**) Principle of DTF detection with a bi-enzymatic biosensor based on laccase and tyrosinase. (**B**) Magnitude of cathodic currents recorded with the bi-enzymatic biosensor for the reduction of p-benzoquinone to p-hydroquinone by square wave voltammetry, reported in [79]. Reproduced fro [79] with permission.

The recovery results from fungicide-spiked vegetables and citrus fruit determined with the biosensors reported in [77,79] were in the range 90.2–101.1%, emphasizing the accuracy of the biosensors and supporting the feasibility of inhibition-based biosensors in detecting ziram and other pesticides in real samples.

Aldehyde dehydrogenase (E.C. 1.2.15, AlDH) catalyzes the transformation of aldehydes to the corresponding carboxylic acids in the presence of the enzymatic cofactor nicotinamide adenine dinucleotide (NAD^+) or nicotinamide adenine dinucleotide phosphate ($NADP^+$):

$$\text{Aldehyde} + NAD^+ + H_2O \quad \overset{\text{ALDH}}{\rightleftharpoons} \quad \text{Carboxylic Acid} + NADH + H^+$$

In the human body, aldehyde dehydrogenase, more specifically ALDH2 is behind the connection between the exposure to pesticides and Parkinson's disease. Aldehyde dehydrogenases, in general, are inhibited by a whole range of compounds, including dithiocarbamate fungicides, benzimidazole fungicides and some heavy metals [88].

In biosensors, measuring the activity of aldehyde dehydrogenase is achieved by determining the reduced cofactor, NADH, formed in the enzymatic reaction. DTFs inhibit the activity of ALDH, causing a decrease in the amount of NADH. The reduced cofactor can be sensitively detected by electrochemical oxidation at the surface of carbon electrodes. Furthermore, the electrochemical detection of NADH is efficiently accomplished using electrochemical mediators or various carbon nanomaterials such as carbon nanotubes, graphene, nanofibers, etc., which decrease the potential required for NADH oxidation [89]. By measuring at low potentials, close to 0 V, the risk of potential interferences from other electrochemically active compounds in samples is minimal, supporting the accuracy of the detection.

Noguer et al. and others have described several biosensors for the detection of dithiocarbamate fungicides, based on the inhibition of aldehyde dehydrogenase from baker's yeast [78,80–82,90]. The equilibrium of the enzymatic reaction favors the reactants' side, consequently, high concentrations of cofactor, alkaline pH, or coupled enzymatic reactions are used to shift the equilibrium of the enzymatic reaction towards the products side and achieve quantitative conversion of aldehydes and the NAD^+ cofactor. A successful approach for the highly sensitive detection of fungicides, such as maneb and zineb [81,82], was to use a second enzyme, either NADH oxidase or diaphorase in order to convert very fast NADH, back to the oxidized form NAD^+.

For example, the determination of maneb and zineb was achieved with a bi-enzyme biosensor comprising aldehyde dehydrogenase and diaphorase [90], according to the following scheme:

$$\text{Propionaldehyde} + \text{NAD}^+ + \text{H}_2\text{O} \overset{\text{ALDH}}{\rightleftharpoons} \text{Acetic acid} + \text{NADH} + \text{H}^+ \tag{1}$$

$$\text{NADH} + [\text{Fe(CN)}_6]^{3-} \xrightarrow{\text{Diaphorase}} \text{NAD}^+ + [\text{Fe(CN)}_6]^{4-} + \text{H}^+ \tag{2}$$

$$[\text{Fe(CN)}_6]^{4-} \xrightarrow{E > E0\prime} [\text{Fe(CN)}_6]^{3-} + e^- \tag{3}$$

The NADH produced in the enzymatic reaction catalyzed by aldehyde dehydrogenase (1) is oxidized back to NAD$^+$ by diaphorase, in a reaction where ferricyanide acts as electron acceptor (2). (3), the ferrocyanide formed in reaction (2) is oxidized at the surface of a Pt electrode, polarized at +0.1 V. The magnitude of the anodic current from (3) is proportional to the activity of aldehyde dehydrogenase in the reaction medium.

The enzymes were used either free in solution or immobilized in a matrix of PVA-SbQ, being retained at the surface of a Pt electrode with a cellophane membrane (Figure 7A). Enzyme immobilization in PVA-SbQ proved advantageous: after incubation with the pesticide solution for 10 min, the biosensor enabled detection as low as 1.48 ppb (maneb) and 9 ppb (zineb) [90] (Figure 7B). It should be highlighted here the approach used to solubilize zineb in alkaline medium in the presence of EDTA, thus transforming it into a disodium salt, known as nabam. Nabam is another DTF that inhibits aldehyde dehydrogenase.

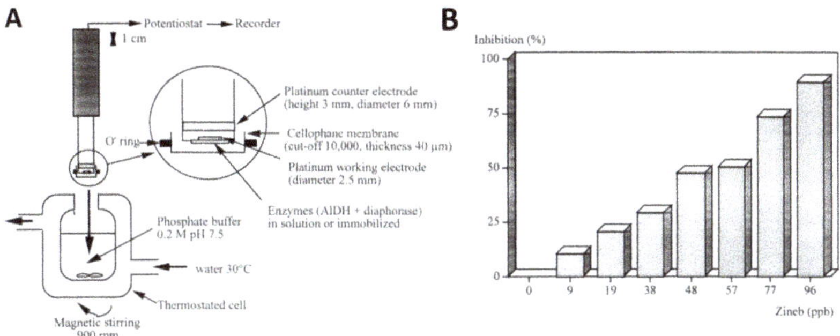

Figure 7. (**A**). Experimental setup used for the bi-enzymatic sensor for maneb and zineb. (**B**). Response of the biosensor at different concentration of zineb, expressed as percentage of the signal in the absence of the inhibitor. Reproduced from [90], by permission.

The use of more stable NADH oxidase instead of diaphorase and disposable screen-printed electrodes instead of Pt electrodes were reported to simplify and reduce the costs of the enzymatic biosensor [82].

While achieving limits of detection in the ppb range, these devices were demonstrated exclusively with standard solutions of pesticides in buffer, with no testing of real samples.

The biosensors based on the inhibition of aldehyde dehydrogenase were developed more than 13 years ago. Re-starting the research efforts in this direction can bring improved performances and wider applications, considering the growing use of nanomaterials in enzymatic inhibition-based biosensors in the last years [75,91] and the opportunities brought by novel enzymes, isolated from various sources. These enzymes have potentially different substrate specificities and inhibitor profiles (discussed below).

The critical factors affecting the analytical performance of the inhibition based enzymatic biosensors are: the amount of enzyme, the incubation time, the design of the device and of the sensing layer, including the matrix and type of immobilization of the enzyme. In general, as summarized in Table 3, the enzymes were entrapped in polymers by photopolymerization (e.g., in PVA-SbQ), adsorbed on electrochemically deposited films of Prussian Blue or were cross-linked with glutaraldehyde in a matrix of bovine serum albumin. The incubation time with the fungicide was less than 20 min. The design of the test must consider the reversibility and the mechanism of inhibition. For example, aldehyde dehydrogenase inhibition by dithiocarbamate fungicides is competitive versus the cofactor NAD$^+$ and is irreversible. Therefore, the analytical signal is registered first in the absence of the fungicide, to get a reference signal, then the NAD$^+$ is eliminated by washing and the signal is recorded in the absence of the cofactor. Next, the biosensor is incubated with the fungicide, followed by a short incubation with NAD$^+$ and by the addition of the substrate, before recording the final signal. The biosensor is discarded after this, since the original activity of the enzyme cannot be restored (Figure 8).

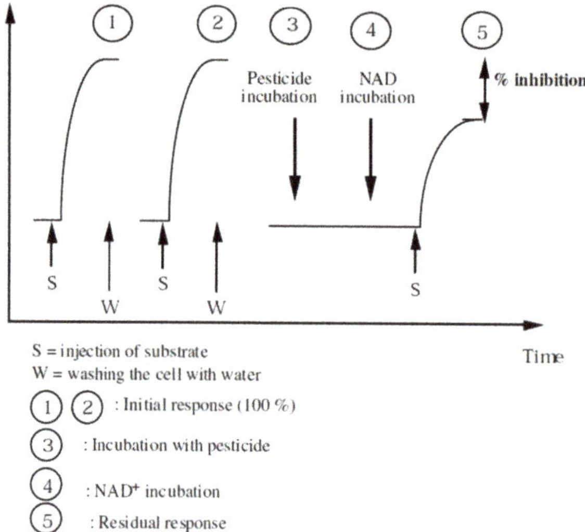

Figure 8. Typical experimental protocol for an amperometric biosensor for DTF based on aldehyde dehydrogenase. Reproduced from [78], by permission.

These steps require an adequate operational stability of the sensor and appear to be too complex to be compatible with in-field testing. A possible solution to circumvent such complexity can be the use of magnetic beads with immobilized enzymes. A magnetic field is applied to "immobilize" the enzyme at electrode surface [92], the enzyme-modified magnetic particles are discarded after the test and the electrode is used for the next measurements.

From a practical application point of view, commercializing such biosensors would require adequate storage and operational stability and production by methods compatible with mass-production, simple use and relatively short analysis time.

With regards to the storage stability, enzymes from extremophilic sources, able to operate in a wider range of temperatures than their mesophilic counterparts, are a viable choice as specific receptors in the biosensor.

The following section details some examples of extremozymes, most promising with regards to stability and therefore with real potential to meet the requirements to be applied in biosensors for practical applications.

2.4.2. Extremozymes as Potential Biorecognition Elements in Biosensors for DTFs

Extremophilic microorganisms constitute a recently exploited reservoir of enzymes stable under various conditions. These microbes developed a particular proteome presenting specific structural and functional features in order to cope with extreme temperatures, hydrostatic pressure, alkaline and acid pHs, high salinity, and radiations [93]. Therefore, for the last few decades, a series of enzymes (extremozymes) were investigated and used for improved industrial processes. Among extremophiles, thermophilic and hyperthermophilic bacteria and archaea represent the most abundant source of extremely stable enzymes that are also resistant to various chemical agents and extreme pH values [94].

Aldehyde dehydrogenases (ALDH, E.C 1.2.1.3) are one of the important yet little explored biocatalysts in biotechnologies and biosensing [95,96]. The inhibition pattern of these extremozymes was poorly studied, in particular the effect of dithiocarbamate derivates known as ALDH inhibitors on their NAD(P)-dependent activity [88]. Nonetheless, the high stability of these enzymes from thermophilic [97], psychrophilic [98], and halophilic microorganisms [99], over a broad pH and temperature range both in solution and immobilized, recommends them as putative enhanced biocatalysts for commercial operations.

Among this class of enzymes from hyperthermophiles, the bifunctional aldehyde dehydrogenase from the archaeon *Pyrococcus furiosus* appeared to be extremely stable, up to 100 °C, and highly active between pH 9.4 and 10.2 at 80 °C [100], while the heterologous ALDH from the thermoacidophilic archaeon *Sulfolobus solfataricus P2* also stable at high temperatures optimally catalyzed the NAD$^+$-dependent aldehydes oxidation at 70 °C and pH 6.5 [101]. ALDH from the thermophilic and alkaliphilic *Natronomonas pharaonis* isolated from a highly saline soda lakes in Egypt; thus, adapted to hypersaline conditions and high pH, showed an optimal temperature of 60 °C in the presence of 0.25 M NaCl at pH 8 [97]. This extremozyme also showed a high long-term stability, preserving its full activity after 24 h storage in various concentrations of NaCl and a specific activity of ~1 µmol·min^{-1}·mg^{-1} when oxidizing acetaldehyde at 20 °C [97].

Scant information is available so far on cold-active aldehyde dehydrogenases, considering that psychrophilic and psychrotolerant microorganisms constitute an important source of stable enzymes that are highly active at low temperatures. Among important cold-active candidates, the ALDH from the psychrotrophic marine Antarctic *Flavobacterium frigidimaris KUC-1* (formerly *Cytophaga* sp.) presented a broad-range thermostability up to 60 °C, with optimal pH >10 [98]. This extremozyme preserved 70% of the activity when incubating at 45 °C for 2 h and has a half-like of 65 min at 50 °C, being showed a lower activation energy at 30 °C as compared to the mesophilic *Saccharomyces cerevisiae* ALDH [102], favoring catalysis at commonly used temperatures in applicative reactions [98].

Moreover, the homologous recombinant enzyme from the same family originating from the Antarctic *Flavobacterium PL002* strain was recently used in an electrochemical test and in a biosensor for the detection of benzaldehyde [89,103]. The enzyme has wide substrate specificity and was shown to be inhibited by thiram [104], thus it appears to be a good candidate as a biorecognition element in a biosensor for DTFs. Moreover, the cold-active PL002 ALDH had only 20% activity reduction after storage for 1 week at 4 °C (C. Purcarea, unpublished data), as an important advantage for biosensing.

Studies of *Rhodococcus sp. N186/21* able to degrade the thiocarbamate herbicide S-ethyl dipropylcarbamothioate revealed the presence of a NAD$^+$-dependent ALDH active on aliphatic aldehydes involved in this cytochrome P-450 related process [105]. Knowing that *Rhodococcus* genus contained widespread polyextremophilic actinobacteria able to survive within a 4 °C to 45 °C temperature range, high hydrostatic pressure, UV irradiation and osmotic stress [106,107], with an extended array of enzymes as putative candidates for environmental and biotechnological applications [108], further investigation of ALDHs from extremophilic *Rhodococcus* species could lead to developments in biosensing for pesticides detection.

Microbial laccases (E.C. 1.10.3.2) constitute currently used sensing biocomponents for pesticides detection [85,109]. During the last decades, the characteristics of a large variety of native and recombinant laccases from thermophilic, psychrophilic, and alkaliphilic bacteria and fungi were reported [110]. The enzyme from the alkalotolerant gammaproteobacterium JB isolated from industrial waste water, optimally active at 55 °C and pH 6.5, was 100% inhibited by 3.5 mM diethyldithiocarbamate when using syringaldazine as substrate, with a Ki of 0.163 mM, an effect reversed by 1.5 mM CuCl$_2$ [111]. Meanwhile, in the case of the thermostable laccase from *Streptomyces lavendulae REN-7*, only a slight inhibition (14%) by sodium N,N-diethyldithiocarbamate trihydrate was observed [112].

Although the inhibitory effect of dithiocarbamate derivates was not investigated for a large variety of these extremozymes, functional studies of laccases from extremophilic microorganisms revealed their high thermal stability and a wide pH interval for catalysis that makes them good candidates for enhanced biosensing components for pesticides monitoring.

2.4.3. Challenges in the Application of Biosensors for DTFs Determination in Real Samples

Most challenges to be solved in order to apply the biosensors for the determination of DTFs in real samples are common with those observed for other pesticides [113], namely, (i) improving the sensitivity, storage stability, reproducibility, and robustness of the sensing device, (ii) demonstration of selectivity in complex matrices, (iii) simplification of sample pre-treatment, and (iv) development of adequate working protocols to allow real time, on field analysis.

An additional hurdle though compared to other types of pesticides that were preferentially studied, is the lack of specific bioreceptors such as antibodies or aptamers and the lack of sample extraction cartridges based on molecularly imprinted polymers [114].

With regards to the sensitivity of enzymatic biosensors, bi-enzymatic devices, such as those based on laccase/tyrosinase and those combining aldehyde dehydrogenase with NADH oxidase or diaphorase, lead to enhanced performance for DTFs detection compared to mono-enzymatic ones. However, complexity, costs, and finding operational conditions that represent a good compromise for both enzymes have to be weighed against the increase in sensitivity. Rational biosensor design, including the controlled immobilization of enzymes and the use of well-characterized nanomaterials and modifiers provides a wealth of possibilities for biosensors for pesticides with improved characteristics [75,91,115].

As with all enzyme-inhibition based biosensors, the selectivity has to be accurately evaluated in accordance to the targeted application. For example, the sensitivity of laccase, tyrosinase and aldehyde dehydrogenase enzymes to several DTFs and other inhibitors can be exploited to develop screening-type systems, alerting on possibly contaminated samples that should be analyzed further by standard methods. Alternatively, sensors arrays such as bioelectronic tongues including several enzymes with different susceptibilities to DTFs, coupled with chemometrics for data analysis can be envisaged for the selective detection of specific compounds, similar to other devices described in literature [116].

Use of enzymes with improved stability, genetically modified enzymes or newly discovered enzymes with different inhibition profiles, inclusion of nanomaterials and stabilizers and fabrication by methods compatible with mass-production are expected to solve issues related to sensitivity, stability, reproducibility and robustness.

The low solubility of DTFs in water is a further complication for sample preparation as are the extraction steps needed to ensure quantitative recovery from food matrices. The development of portable biosensor-based devices can be envisaged however, for the fast screening of DTFs on the surface of intact fruits and vegetables and wherever DTFs can be converted easily to soluble salts. In this sense, a wearable glove biosensor, based on the inhibition of acetylcholinesterase that was recently demonstrated for the detection of organophosphorus pesticides on the surface of intact vegetables [117] can serve as a model.

Comparison with standard confirmatory methods, lacking in biosensor reports will go a long way to support the feasibility of such devices and is anticipated to encourage research in this direction.

3. Conclusions and Perspectives

Despite the large use of DTFs in agriculture, research in the field of analytical methods dedicated to these fungicides has been very limited compared to other classes of pesticides. Today, like 20 years ago [72], most standard methods for the analysis of DTFs remain based on the degradation of fungicides to CS_2 and measuring the resulting amount of CS_2, which does not make it possible to discriminate between various DTFs with different toxicities.

A small number of chromatographic methods coupled with mass spectrometric detection enable the separation and detection of DTFs from different groups, i.e., propylene-bis-, ethylene-bis and dimethyl-dithiocarbamates. Moreover the specific detection of metal containing DTFs was achieved by coupling chromatographic separation with atomic absorption spectroscopy. Recent advances in chromatography based approaches mainly concerned sample extraction procedures.

Most research efforts in the last years were concentrated on the SERS-based detection of DTFs, targeting: (i) portable devices, (ii); simple and effective sampling procedures; and above all (iii) the discovery of new cost-effective Ag, Au, and Cu-based nanomaterials and composites as SERS substrates for the highly sensitive detection of DTFs and thiram in particular. Various nanomaterial morphologies, modifiers for specific anchoring the analyte to the hot spots, 2D and hierarchical 3D nanostructures have been investigated aiming for high sensing area with a high density of "hot spots" for enhanced signals. In line with the current trends, significant advances were noted for the "paste and peel" sensors based on flexible supports, which could be brought in close contact with the sample surface and thus play a double role, i.e., sampling and sensing, to enable the fast, in situ analysis of thiram. The detection of thiram, as a prominent example of DTFs was demonstrated in a variety of spiked samples, including soil, strawberries, tomato, cucumber, water, etc., for which satisfactory recoveries were calculated. Without any doubt, considering the effervescence of research on this topic, the progress with portable devices and the efforts towards enhancing the accuracy of the detection (e.g., by including internal standards, using chemometrics for analyzing the data etc.), SERS-based methods have a very high potential to achieve in situ selective detection of specific DTFs.

In addition to SERS, many assays based on optical and electrochemical detection have been developed. Oftentimes the accent was placed on developing simple and low cost procedures. However, the selectivity of most devices was not unambiguously demonstrated, and their accuracy remains to be confirmed by comparison with standard methods.

While biosensors appear as an attractive alternative to separation-based methods with their promise for fast, selective, cost effective, portable, and simple detection, the detection of DTFs was rarely explored. This is in part because antibodies, aptamers, and molecularly imprinted polymers for DTFs are not available commercially as for other pesticides. The few reports on biosensors for DTFs based on the inhibition of laccase, tyrosinase or aldehyde dehydrogenase (some more than 10 years old) emphasize their capability to reach detection limits compatible with practical applications. The development of biosensors for DTFs can be fast tracked by exploiting the knowledge and adapting concepts from biosensors for other pesticides, e.g., the widely studied organophosphates and carbamates that inhibit cholinesterases.

More specifically, controlled immobilization of enzymes, use of nanomaterials to enhance the electrochemical signal, embracing the trend for flexible devices used for both sampling and detection can potentially lead to highly sensitive biosensors for DTFs for in situ detection applications. For example, many opportunities are anticipated for devices such as biosensor swipes or gloves for thiram detection on the surface of fruits and vegetables.

Considering that laccase, tyrosinase and aldehyde dehydrogenase are inhibited in different proportions not by a single fungicide, but by a group of DTFs as well as by several other compounds, the most direct application of enzyme-based biosensors is as alert systems. Selective detection of a particular compound might be attempted in the future with bioelectronic tongues. Enzymes obtained by engineering approaches, with improved selectivity and stability, or new enzymes isolated from extremophiles, with new substrate specificity profile and enhanced stability can contribute to meet the requirements for practical applications of biosensors in DTFs analysis.

Obviously, there is a high need of stable, specific receptors such as aptamers or molecularly imprinted polymers that could simplify the analysis of DTFs and assist not only with detection but also with sample extraction and cleaning. Two main strategies are currently pursued with sample pre-treatment: development of QuEChERS methods, suitable for laboratory-based analysis of any type of sample and fast methods, such as "paste and peel","swipe", etc., mainly intended for fast, in situ sampling, resulting in flexible films adhering to sample surfaces that collect the contaminants and are afterwards directly used for sensing by SERS. While acknowledging limitations due to the low solubility in water and organic solvents of most DTFs, further progress enabling fast, quantitative recovery of DTFs is expected for specific applications, such as the detection of thiram from the surface of intact fruits and vegetables.

In summary, although various approaches are available for the detection of DTFs, selectivity remains a critical issue to be addressed in a more detailed and application-oriented manner in the coming years. Clearly, there are many analytical opportunities ahead in the analysis of DTFs and the field is one deserving far more concentrated research efforts.

Author Contributions: Conceptualization, P.F.-B., J.L., C.P. and A.V.; methodology, P.F.-B., J.L. and A.V.; investigation, P.F.-B., J.L., R.F., C.P. and A.V.; writing—original draft preparation, P.F.-B., J.L., R.F., C.P. and A.V.; writing—review and editing, J.L. and A.V. All authors have read and agreed to the published version of the manuscript.

Funding: This research was funded through the MEraNet grant "ENZ4IFACES" by the Romanian Executive Agency for Higher Education, Research, Development and Innovation (UEFISCDI), contract 166/1.05.2020 (for C.P and A.V), Asturias-Subvenciones dirigidas a empresas (IDEPA), Spain (for P.F.B) and the Department of Science and Innovation (DSI) of South Africa (for J.L. and R.F).

Institutional Review Board Statement: Not applicable.

Informed Consent Statement: Not applicable.

Data Availability Statement: No new data were created or analyzed in this study. Data sharing is not applicable to this article.

Conflicts of Interest: The authors declare no conflict of interest.

References

1. Lefton, J.B.; Pekar, K.B.; Runčevski, T. The Crystal Structure of Zineb, Seventy-Five Years Later. *Cryst. Growth Des.* **2020**, *20*, 851–857. [CrossRef]
2. Fungicides Market Report—Global Industry Analysis, Size, Share, Growth, Trends, and Forecast 2017–2025. Available online: https://www.prnewswire.com/news-releases/fungicides-market-mancozeb-chlorothalonil-metalaxyl-strobilurin-and-others-for-cereals--grains-oilseeds--pulses-fruits--vegetables-and-other-crops---global-industry-analysis-size-share-growth-trends-and-forecast-20-300243678.html (accessed on 29 November 2020).
3. European Commission Regulation 2016/1 of 3 December 2015 Amending Annexes II and III to Regulation (EC) No 396/2005 of the European Parliament and of the Council as Regards Maximum Residue Levels for Bifenazate, Boscalid, Cyazofamid, Cyromazine, Dazomet, Dithiocarbamates, Fluazifop-P, Mepanipyrim, Metrafenone, Picloram, Propamocarb, Pyridaben, Pyriofenone, Sulfoxaflor, Tebuconazole, Tebufenpyrad and Thiram in or on Certain Products. Available online: https://eur-lex.europa.eu/eli/reg/2016/1/oj (accessed on 29 November 2020).
4. Atuhaire, A.; Kaye, E.; Mutambuze, I.L.; Matthews, G.; Friedrich, T.; Jørs, E. Assessment of Dithiocarbamate Residues on Tomatoes Conventionally Grown in Uganda and the Effect of Simple Washing to Reduce Exposure Risk to Consumers. *Environ. Health Insights* **2017**, *11*. [CrossRef] [PubMed]
5. European Food Safety Agency. The 2017 European Union report on pesticide residues in food. *EFSA J.* **2019**, *17*, e05743.
6. Nakamura, M.; Noda, S.; Kosugi, M.; Ishiduka, N.; Mizukoshi, K.; Taniguchi, M.; Nemoto, S. Determination of dithiocarbamates and milneb residues in foods by gas chromatography-mass spectrometry. *J. Food Hyg. Soc. Jpn.* **2010**, *51*, 213–219. [CrossRef]
7. Crnogorac, G.; Schmauder, S.; Schwack, W. Trace analysis of dithiocarbamate fungicide residues on fruits and vegetables by hydrophilic interaction liquid chromatography/tandem mass spectrometry. *Rapid Commun. Mass Spectrom.* **2008**, *22*, 2539–2546. [CrossRef]
8. Kakitani, A.; Yoshioka, T.; Nagatomi, Y.; Harayama, K. A rapid and sensitive analysis of dithiocarbamate fungicides using modified QuEChERS method and liquid chromatography-tandem mass spectrometry. *J. Pestic. Sci.* **2017**, *42*, 145–150. [CrossRef]
9. Li, J.; Dong, C.; Yang, Q.; An, W.; Zheng, Z.; Jiao, B. Simultaneous Determination of Ethylenebisdithiocarbamate (EBDC) and Propylenebisdithiocarbamate (PBDC) Fungicides in Vegetables, Fruits, and Mushrooms by Ultra-High-Performance Liquid Chromatography Tandem Mass Spectrometry. *Food Anal. Methods* **2019**, *12*, 2045–2055. [CrossRef]
10. Van Lishaut, H.; Schwack, W. Selective trace determination of dithiocarbamate fungicides in fruits and vegetables by reversed-phase ion pair liquid chromatography with ultraviolet and electrochemical detection. *J. AOAC Int.* **2000**, *83*, 720–727. [CrossRef]

11. Al-Alam, J.; Bom, L.; Chbani, A.; Fajloun, Z.; Millet, M. Analysis of Dithiocarbamate Fungicides in Vegetable Matrices Using HPLC-UV Followed by Atomic Absorption Spectrometry. *J. Chromatogr. Sci.* **2017**, *55*, 429–435. [CrossRef]

12. Ubeda, M.R.; Escribano, M.T.S.; Hernandez, L.H. Determination of Thiram by high-performance liquid chromatography with amperometric detection in river water and fungicide formulations. *Microchem. J.* **1990**, *41*, 22–28. [CrossRef]

13. Shapovalova, E.N.; Yaroslavtseva, L.N.; Merkulova, N.L.; Yashin, A.Y.; Shpigun, O.A. Separation of pesticides by high-performance liquid chromatography with amperometric detection. *J. Anal. Chem.* **2009**, *64*, 164–170. [CrossRef]

14. Charoenkitamorn, K.; Chailapakul, O.; Siangproh, W. Development of gold nanoparticles modified screen-printed carbon electrode for the analysis of thiram, disulfiram and their derivative in food using ultra-high performance liquid chromatography. *Talanta* **2015**, *132*, 416–423. [CrossRef] [PubMed]

15. Rahimi, F.; Chatzimichail, S.; Saifuddin, A.; Surman, A.J.; Taylor-Robinson, S.D.; Salehi-Reyhani, A. A Review of Portable High-Performance Liquid Chromatography: The Future of the Field? *Chromatographia* **2020**, *83*, 1165–1195. [CrossRef]

16. Fleischmann, M.; Hendra, P.J.; McQuillan, A.J. Raman spectra of pyridine adsorbed at a silver electrode. *Chem. Phys. Lett.* **1974**, *26*, 163–166. [CrossRef]

17. Martín-Yerga, D.; Pérez-Junquera, A.; González-García, M.B.; Hernández-Santos, D.; Fanjul-Bolado, P. Towards single-molecule in situ electrochemical SERS detection with disposable substrates. *Chem. Commun.* **2018**, *54*, 5748–5751. [CrossRef]

18. Mylrajan, M. SERS, FT-Raman and FT-IR studies of dithiocarbamates. *J. Mol. Struct.* **1995**, *348*, 257–260. [CrossRef]

19. Tse Yuen, K. SERS of dithiocarbamates and xanthates. *Spectrochim. Acta A Mol. Biomol. Spectrosc.* **1995**, *51*, 2177–2192. [CrossRef]

20. Yuan, C.; Liu, R.; Wang, S.; Han, G.; Han, M.-Y.; Jiang, C.; Zhang, Z. Single clusters of self-assembled silver nanoparticles for surface-enhanced Raman scattering sensing of a dithiocarbamate fungicide. *J. Mat. Chem.* **2011**, *21*, 16264–16270. [CrossRef]

21. Zhu, C.; Wang, X.; Shi, X.; Yang, F.; Meng, G.; Xiong, Q.; Ke, Y.; Wang, H.; Lu, Y.; Wu, N. Detection of Dithiocarbamate Pesticides with a Spongelike Surface-Enhanced Raman Scattering Substrate Made of Reduced Graphene Oxide-Wrapped Silver Nanocubes. *ACS Appl. Mater. Interfaces* **2017**, *9*, 39618–39625. [CrossRef]

22. Nowicka, A.B.; Czaplicka, M.; Kowalska, A.A.; Szymborski, T.; Kamińska, A. Flexible PET/ITO/Ag SERS Platform for Label-Free Detection of Pesticides. *Biosensors* **2019**, *9*, 111. [CrossRef]

23. Kumar, G.; Soni, R.K. Silver Nanocube- and Nanowire-Based SERS Substrates for Ultra-low Detection of PATP and Thiram Molecules. *Plasmonics* **2020**, *15*, 1577–1589. [CrossRef]

24. Sun, L.; Wang, C. Highly Sensitive and Rapid Surface Enhanced Raman Spectroscopic (SERS) Determination of Thiram on the Epidermis of Fruits and Vegetables Using A Silver Nanoparticle-Modified Fibrous Swab. *Anal. Lett.* **2020**, *53*, 973–983. [CrossRef]

25. Saute, B.; Premasiri, R.; Ziegler, L.; Narayanan, R. Gold nanorods as surface enhanced Raman spectroscopy substrates for sensitive and selective detection of ultra-low levels of dithiocarbamate pesticides. *Analyst* **2012**, *137*, 5082–5087. [CrossRef] [PubMed]

26. Saute, B.; Narayanan, R. Solution-based SERS method to detect dithiocarbamate fungicides in different real-world matrices. *J. Raman Spectrosc.* **2013**, *44*, 1518–1522. [CrossRef]

27. Zhao, Y.; Pérez-Segarra, W.; Shi, Q.; Wei, A. Dithiocarbamate Assembly on Gold. *JACS* **2005**, *127*, 7328–7329. [CrossRef]

28. Chen, M.; Luo, W.; Liu, Q.; Hao, N.; Zhu, Y.; Liu, M.; Wang, L.; Yang, H.; Chen, X. Simultaneous In Situ Extraction and Fabrication of Surface-Enhanced Raman Scattering Substrate for Reliable Detection of Thiram Residue. *Anal. Chem.* **2018**, *90*, 13647–13654. [CrossRef]

29. Ibáñez, D.; González-García, M.B.; Hernández-Santos, D.; Fanjul-Bolado, P. Detection of dithiocarbamate, chloronicotinyl and organophosphate pesticides by electrochemical activation of SERS features of screen-printed electrodes. *Spectrochim. Acta A* **2020**, 119174. [CrossRef]

30. Cassella, A.R.; Cassella, R.J.; Garrigues, S.; Santelli, R.E.; de Campos, R.C.; de la Guardia, M. Flow injection-FTIR determination of dithiocarbamate pesticides. *Analyst* **2000**, *125*, 1829–1833.

31. Cassella, A.R.; Garrigues, S.; de Campos, R.C.; de la Guardia, M. Fourier transform infrared spectrometric determination of Ziram. *Talanta* **2001**, *54*, 1087–1094. [CrossRef]

32. Halls, D.J. The properties of dithiocarbamates A Review. *Microchim Acta* **1969**, *57*, 62–77. [CrossRef]

33. Bond, A.M.; Martin, R.L. Electrochemistry and redox behaviour of transition metal dithiocarbamates. *Coord. Chem. Rev.* **1984**, *54*, 23–98. [CrossRef]

34. Giannakopoulos, E.; Deligiannakis, Y. Thermodynamics of Adsorption of Dithiocarbamates at the Hanging Mercury Drop. *Langmuir* **2007**, *23*, 2453–2462. [CrossRef] [PubMed]

35. Labuda, J.; Mocák, J.; Bustin, D.I. Electrochemical study of the diethyldithiocarbamate anion and of its oxidation products. *Chem. Pap.* **1981**, *36*, 633–664.

36. Silva, R.A.G.; Silva, L.A.J.; Munoz, R.A.A.; Richter, E.M.; Oliveira, A.C. Fast and direct determination of mancozeb through batch injection analysis with amperometric detection on boron-doped diamond electrodes. *J. Electroanal. Chem.* **2014**, *733*, 85–90. [CrossRef]

37. Qiu, P.; Ni, Y.N. Determination of ziram in vegetable samples by square wave voltammetry. *Chin. Chem. Lett.* **2008**, *19*, 1337–1340. [CrossRef]

38. Silva, L.M.; De Souza, D. Ziram herbicide determination using a polished silver solid amalgam electrode. *Electrochim. Acta* **2017**, *224*, 541–550. [CrossRef]

39. Procopio, J.R.; Sevilla Escribano, M.T.; Hernandez, L.H. Determination of thiram in water and soils by cathodic stripping voltammetry based on adsorptive accumulation. *Fresenius Z. Anal. Chem.* **1988**, *331*, 27–29. [CrossRef]

40. Sevilla, M.T.; Procopio, J.R.; Pinilla, J.M.; Hernandez, L. Voltammetric determination of thiram following adsorptive accumulation on a rotating gold disk electrode. *Electroanalysis* **1990**, *2*, 475–479. [CrossRef]
41. Fernández, C.; Reviejo, A.J.; Pingarrón, J.M. Development of graphite-poly(tetrafluoroethylene) composite electrodes Voltammetric determination of the herbicides thiram and disulfiram. *Anal. Chim. Acta* **1995**, *305*, 192–199. [CrossRef]
42. Fernández, C.; Reviejo, A.J.; Pingarrón, J.M. Graphite-poly(tetrafluoroethylene) electrodes as electrochemical detectors in flowing systems. *Anal. Chim. Acta* **1995**, *314*, 13–22. [CrossRef]
43. Mathew, L.; Reddy, M.L.P.; Rao, T.P.; Iyer, C.S.P.; Damodaran, A.D. Differential pulse anodic stripping voltammetric determination of ziram (a dithiocarbamate fungicide). *Talanta* **1996**, *43*, 73–76. [CrossRef]
44. Shan Lin, M.; Iuan Jan, B.; Leu, H.-J.; Shing Lin, J. Trace measurement of dithiocarbamate based pesticide by adsorptive stripping voltammetry. *Anal. Chim. Acta* **1999**, *388*, 111–117. [CrossRef]
45. Shaidarova, L.G.; Budnikov, G.K.; Zaripova, S.A. Electrocatalytic Determination of Dithiocarbamate-Based Pesticides Using Electrodes Modified with Metal Phthalocyanines. *J. Anal. Chem.* **2001**, *56*, 748–753. [CrossRef]
46. Lin, M.S.; Wang, J.S. Determination of an Ethylene Bisdithiocarbamate Based Pesticide (Nabam) by Cobalt Phthalocyanine Modified Carbon Ink Electrode. *Electroanalysis* **2004**, *16*, 904–909. [CrossRef]
47. Zhao, Y.-G.; Zheng, X.-W.; Huang, Z.-Y.; Yang, M.-M. Voltammetric study on the complex of thiram–copper(II) and its application. *Anal. Chim. Acta* **2003**, *482*, 29–36. [CrossRef]
48. Barroso, M.F.; Paíga, P.; Vaz, M.C.V.F.; Delerue-Matos, C. Study of the voltammetric behaviour of metam and its application to an amperometric flow system. *Anal. Bioanal. Chem.* **2005**, *383*, 880–885. [CrossRef] [PubMed]
49. Nováková, K.; Navrátil, T.; Dytrtová, J.J.; Chýlková, J. The use of copper solid amalgam electrodes for determination of the pesticide thiram. *J. Solid State Electrochem.* **2013**, *17*, 1517–1528. [CrossRef]
50. Abbaci, A.; Azzouz, N.; Bouznit, Y. A new copper doped montmorillonite modified carbon paste electrode for propineb detection. *Appl. Clay Sci.* **2014**, *90*, 130–134. [CrossRef]
51. López-Fernández, O.; Barroso, M.F.; Fernandes, D.M.; Rial-Otero, R.; Simal-Gándara, J.; Morais, S.; Nouws, H.P.A.; Freire, C.; Delerue-Matos, C. Voltammetric analysis of mancozeb and its degradation product ethylenethiourea. *J. Electroanal. Chem.* **2015**, *758*, 54–58. [CrossRef]
52. Stanković, D.M.; Kalcher, K. Amperometric quantification of the pesticide ziram at boron doped diamond electrodes using flow injection analysis. *Sens. Actuators B Chem.* **2016**, *233*, 144–147. [CrossRef]
53. Stanković, D.M. Electroanalytical Approach for Quantification of Pesticide Maneb. *Electroanalysis* **2017**, *29*, 352–357. [CrossRef]
54. Nishiyama, K.; Sawada, N.; Kataoka, M.; Tsuruta, K.; Yamamura, K.; Origuchi, S.; Yoshimoto, S. Electrochemical detection of manzeb using reductive desorption from Au(111) and Au(100). *Electrochemistry* **2018**, *86*, 345–348. [CrossRef]
55. Zamora-Sequeira, R.; Alvarado-Hidalgo, F.; Robles-Chaves, D.; Sáenz-Arce, G.; Avendano-Soto, E.D.; Sánchez-Kopper, A.; Starbird-Perez, R. Electrochemical Characterization of Mancozeb Degradation for Wastewater Treatment Using a Sensor Based on Poly (3,4-ethylenedioxythiophene) (PEDOT) Modified with Carbon Nanotubes and Gold Nanoparticles. *Polymers* **2019**, *11*, 1449. [CrossRef] [PubMed]
56. Maximiano, E.M.; Cardoso, C.A.L.; Arruda, G.J. Simultaneous Electroanalytical Determination of Thiram and Carbendazim in Samples of Fresh Fruit Juices in the Presence of Surfactants. *Food Anal. Methods* **2020**, *13*, 119–130. [CrossRef]
57. Ragam, P.N.; Mathew, B. Unmodified silver nanoparticles for dual detection of dithiocarbamate fungicide and rapid degradation of water pollutants. *Int. J. Environ. Sci. Technol.* **2020**, *17*, 1739–1752. [CrossRef]
58. Alves Sá da Silva, V.; da Silva Santos, A.; Ferreira, T.L.; Codognoto, L.; Agostini Valle, E.M. Electrochemical Evaluation of Pollutants in the Environment: Interaction Between the Metal Ions Zn(II) and Cu(II) with the Fungicide Thiram in Billings Dam. *Electroanalysis* **2020**, *32*, 1582–1589. [CrossRef]
59. Da Silva, M.d.P.; Procopio, J.R.; Hernández, L. Electrochemical detection in the determination of several dithiocarbamates by reverse-phase liquid chromatography. *J. Liq. Chromatogr. Rel. Technol.* **1999**, *22*, 463–475. [CrossRef]
60. Fernández, C.; Reviejo, A.J.; Polo, L.M.; Pingarrón, J. HPLC-Electrochemical detection with graphite-poly (tetrafluoroethylene) electrode: Determination of the fungicides thiram and disulfiram. *Talanta* **1996**, *4*, 1341–1348. [CrossRef]
61. Zagal, J.H.; Griveau, S.; Silva, J.F.; Nyokong, T.; Bedioui, F. Metallophthalocyanine-based molecular materials as catalysts for electrochemical reactions. *Coord. Chem. Rev.* **2010**, *254*, 2755–2791. [CrossRef]
62. Syafiuddin, A.; Salmiati; Salim, M.R.; Beng Hong Kueh, A.; Hadibarata, T.; Nur, H. A Review of Silver Nanoparticles: Research Trends, Global Consumption, Synthesis, Properties, and Future Challenges. *J. Chin. Chem. Soc.* **2017**, *64*, 732–756. [CrossRef]
63. Ghoto, S.A.; Khuhawar, M.Y.; Jahangir, T.M. Applications of copper nanoparticles for colorimetric detection of dithiocarbamate pesticides. *J. Nanostructure Chem.* **2019**, *9*, 77–93. [CrossRef]
64. Ghoto, S.A.; Khuhawar, M.Y.; Jahangir, T.M. Silver Nanoparticles with Sodium Dodecyl Sulfate as a Colorimetric Probe for the Detection of Dithiocarbamate Pesticides in Environmental Samples. *Anal. Sci.* **2019**, *35*, 631–637. [CrossRef] [PubMed]
65. Giannoulis, K.M.; Giokas, D.L.; Tsogas, G.Z.; Vlessidis, A.G. Ligand-free gold nanoparticles as colorimetric probes for the non-destructive determination of total dithiocarbamate pesticides after solid phase extraction. *Talanta* **2014**, *119*, 276–283. [CrossRef] [PubMed]
66. Wang, D.; Liu, D.; Duan, H.; Xu, Y.; Zhou, Z.; Wang, P. Catechol Dyes–Tyrosinase System for Colorimetric Determination and Discrimination of Dithiocarbamate Pesticides. *J. Agric. Food Chem.* **2020**, *68*, 9252–9259. [CrossRef]

67. Yang, L.; Zhang, X.; Wang, J.; Sun, H.; Jiang, L. Double-decrease of the fluorescence of CdSe/ZnS quantum dots for the detection of zinc(II) dimethyldithiocarbamate (ziram) based on its interaction with gold nanoparticles. *Microchim. Acta* **2018**, *185*, 472. [CrossRef]

68. Zhang, C.; Zhang, K.; Zhao, T.; Liu, B.; Wang, Z.; Zhang, Z. Selective phosphorescence sensing of pesticide based on the inhibition of silver(I) quenched ZnS:Mn2+ quantum dots. *Sens. Actuators B Chem.* **2017**, *252*, 1083–1088. [CrossRef]

69. Waseem, A.; Yaqoob, M.; Nabi, A. Determination of thiram in natural waters using flow-injection with cerium(IV)–quinine chemiluminescence system. *Luminescence* **2010**, *25*, 71–75. [CrossRef]

70. Asghar, M.; Yaqoob, M.; Haque, N.; Nabi, A. Determination of Thiram and Aminocarb Pesticides in Natural Water Samples Using Flow Injection with Tris(2,2′-bipyridyl)ruthenium(II)-diperiodatoargentate(III) Chemiluminescence Detection. *Anal. Sci.* **2013**, *29*, 1061–1066. [CrossRef]

71. Girotti, S.; Maiolini, E.; Ghini, S.; Ferri, E.; Fini, F.; Nodet, P.; Eremin, S. Quantification of Thiram in Honeybees: Development of a Chemiluminescent ELISA. *Anal. Lett.* **2008**, *41*, 46–55. [CrossRef]

72. Queffelec, A.-L.; Boisdé, F.; Larue, J.-P.; Haelters, J.-P.; Corbel, B.; Thouvenot, D.; Nodet, P. Development of an Immunoassay (ELISA) for the Quantification of Thiram in Lettuce. *J. Agric. Food Chem.* **2001**, *49*, 1675–1680. [CrossRef]

73. Van Dyk, J.S.; Pletschke, B. Review on the use of enzymes for the detection of organochlorine, organophosphate and carbamate pesticides in the environment. *Chemosphere* **2011**, *82*, 291–307. [CrossRef] [PubMed]

74. Bucur, B.; Munteanu, F.-D.; Marty, J.-L.; Vasilescu, A. Advances in Enzyme-Based Biosensors for Pesticide Detection. *Biosensors* **2018**, *8*, 27. [CrossRef] [PubMed]

75. Kurbanoglu, S.; Ozkan, S.A.; Merkoçi, A. Nanomaterials-based enzyme electrochemical biosensors operating through inhibition for biosensing applications. *Biosens. Bioelectron.* **2017**, *89*, 886–898. [CrossRef] [PubMed]

76. Arduini, F.; Amine, A. Biosensors based on enzyme inhibition. In *Advances in Biochemical Engineering/Biotechnology*; Springer: Berlin/Heidelberg, Germany, 2014; Volume 140, pp. 299–326.

77. Oliveira, T.M.B.F.; Fátima Barroso, M.; Morais, S.; Araújo, M.; Freire, C.; de Lima-Neto, P.; Correia, A.N.; Oliveira, M.B.P.P.; Delerue-Matos, C. Laccase-Prussian blue film-graphene doped carbon paste modified electrode for carbamate pesticides quantification. *Biosens. Bioelectron.* **2013**, *47*, 292–299. [CrossRef]

78. Noguer, T.; Balasoiu, A.-M.; Vasilescu, A.; Marty, J. Development of a disposable biosensor for the detection of metam-sodium and its metabolite MITC. *Anal. Lett.* **2001**, *34*, 513–528. [CrossRef]

79. Oliveira, T.; Barroso, M.F.; Morais, S.; Araújo, M.; Freire, C.; Lima-Neto, P.; Correia, A.; Oliveira, M.; Delerue-Matos, C. Sensitive bi-enzymatic biosensor based on polyphenoloxidases-gold nanoparticles-chitosan hybrid film-graphene doped carbon paste electrode for carbamates detection. *Bioelectrochemistry* **2014**, *98C*, 20–29. [CrossRef]

80. Lima, R.S.; Nunes, G.S.; Noguer, T.; Marty, J.-L. Biossensor enzimático para detecção de fungicidas ditiocarbamatos: Estudo cinético da enzima aldeído desidrogenase e otimização do biossensor. *Quim. Nova* **2007**, *30*, 9–17. [CrossRef]

81. Noguer, T.; Marty, J.-L. High sensitive bienzymic sensor for the detection of dithiocarbamate fungicides. *Anal. Chim. Acta* **1997**, *347*, 63–70. [CrossRef]

82. Noguer, T.; Gradinaru, A.; Ciucu, A.; Marty, J. A New Disposable Biosensor for the Accurate and Sensitive Detection of Ethylenebis(Dithiocarbamate) Fungicides. *Anal. Lett.* **1999**, *32*, 1723–1738. [CrossRef]

83. Flampouri, K.; Mavrikou, S.; Kintzios, S.; Miliadis, G.; Aplada-Sarlis, P. Development and validation of a cellular biosensor detecting pesticide residues in tomatoes. *Talanta* **2010**, *80*, 1799–1804. [CrossRef]

84. Luque de Castro, M.D.; Herrera, M.C. Enzyme inhibition-based biosensors and biosensing systems: Questionable analytical devices. *Biosens. Bioelectron.* **2003**, *18*, 279–294. [CrossRef]

85. Zapp, E.; Brondani, D.; Vieira, I.C.; Scheeren, C.W.; Dupont, J.; Barbosa, A.M.J.; Ferreira, V.S. Biomonitoring of methomyl pesticide by laccase inhibition on sensor containing platinum nanoparticles in ionic liquid phase supported in montmorillonite. *Sens. Actuators B Chem.* **2011**, *155*, 331–339. [CrossRef]

86. Oliveira, T.M.B.F.; Fátima Barroso, M.; Morais, S.; de Lima-Neto, P.; Correia, A.N.; Oliveira, M.B.P.P.; Delerue-Matos, C. Biosensor based on multi-walled carbon nanotubes paste electrode modified with laccase for pirimicarb pesticide quantification. *Talanta* **2013**, *106*, 137–143. [CrossRef]

87. Wang, T.; Milton, R.D.; Abdellaoui, S.; Hickey, D.P.; Minteer, S.D. Laccase Inhibition by Arsenite/Arsenate: Determination of Inhibition Mechanism and Preliminary Application to a Self-Powered Biosensor. *Anal. Chem.* **2016**, *88*, 3243–3248. [CrossRef] [PubMed]

88. Koppaka, V.; Thompson, D.C.; Chen, Y.; Ellermann, M.; Nicolaou, K.C.; Juvonen, R.O.; Petersen, D.; Deitrich, R.A.; Hurley, T.D.; Vasiliou, V. Aldehyde dehydrogenase inhibitors: A comprehensive review of the pharmacology, mechanism of action, substrate specificity, and clinical application. *Pharmacol. Rev.* **2012**, *64*, 520–539. [CrossRef] [PubMed]

89. Titoiu, A.M.; Lapauw, M.; Necula-Petrareanu, G.; Purcarea, C.; Fanjul-Bolado, P.; Marty, J.-L.; Vasilescu, A. Carbon Nanofiber and Meldola Blue Based Electrochemical Sensor for NADH: Application to the Detection of Benzaldehyde. *Electroanalysis* **2018**, *30*, 2676–2688. [CrossRef]

90. Noguer, T.; Leca, B.; Jeanty, G.; Marty, J.-L. Biosensors based on enzyme inhibition: Detection of organophosphorus and carbamate insecticides and dithiocarbamate fungicides. *Field Anal. Chem. Technol.* **1999**, *3*, 171–178. [CrossRef]

91. Arduini, F.; Cinti, S.; Scognamiglio, V.; Moscone, D. Nanomaterials in electrochemical biosensors for pesticide detection: Advances and challenges in food analysis. *Microchim. Acta* **2016**, *183*, 2063–2083. [CrossRef]

92. Llopis, X.; Pumera, M.; Alegret, S.; Merkoçi, A. Lab-on-a-chip for ultrasensitive detection of carbofuran by enzymatic inhibition with replacement of enzyme using magnetic beads. *Lab Chip* **2009**, *9*, 213–218. [CrossRef]

93. Vasudevan, N.; Jayshree, A. Extremozymes and Extremoproteins in Biosensor Applications. In *Encyclopedia of Marine Biotechnology*; Wiley: Hoboken, NJ, USA, 2020; pp. 1711–1736.

94. Vieille, C.; Zeikus, G.J. Hyperthermophilic Enzymes: Sources, Uses, and Molecular Mechanisms for Thermostability. *Microbiol. Mol. Biol. Rev.* **2001**, *65*, 1–43. [CrossRef]

95. Xu, F. Applications of oxidoreductases: Recent progress. *Ind. Biotechnol.* **2005**, *1*, 38–50. [CrossRef]

96. Singh, R.; Kumar, M.; Mittal, A.; Mehta, P.K. Microbial enzymes: Industrial progress in 21st century. *3 Biotech* **2016**, *6*, 174. [CrossRef] [PubMed]

97. Cao, Y.; Liao, L.; Xu, X.W.; Oren, A.; Wu, M. Aldehyde dehydrogenase of the haloalkaliphilic archaeon *Natronomonas pharaonis* and its function in ethanol metabolism. *Extremophiles* **2008**, *12*, 849–854. [CrossRef] [PubMed]

98. Yamanaka, Y.; Kazuoka, T.; Yoshida, M.; Yamanaka, K.; Oikawa, T.; Soda, K. Thermostable aldehyde dehydrogenase from psychrophile, *Cytophaga* sp. KUC-1: Enzymological characteristics and functional properties. *Biochem. Biophys. Res. Commun.* **2002**, *298*, 632–637. [CrossRef]

99. Kim, H.-j.; Joo, W.-A.; Cho, C.-W.; Kim, C.-W. Halophile Aldehyde Dehydrogenase from Halobacterium salinarum. *J. Proteome Res.* **2006**, *5*, 192–195. [CrossRef]

100. Keller, M.W.; Lipscomb, G.L.; Nguyen, D.M.; Crowley, A.T.; Schut, G.J.; Scott, I.; Kelly, R.M.; Adams, M.W.W. Ethanol production by the hyperthermophilic archaeon *Pyrococcus furiosus* by expression of bacterial bifunctional alcohol dehydrogenases. *Microb. Biotechnol.* **2017**, *10*, 1535–1545. [CrossRef]

101. Esser, D.; Kouril, T.; Talfournier, F.; Polkowska, J.; Schrader, T.; Bräsen, C.; Siebers, B. Unraveling the function of paralogs of the aldehyde dehydrogenase super family from *Sulfolobus solfataricus*. *Extremophiles* **2013**, *17*, 205–216. [CrossRef]

102. Tamaki, N.; Hama, T. Aldehyde dehydrogenase from bakers' yeast. *Methods Enzymol.* **1982**, *89*, 469–473.

103. Titoiu, A.M.; Necula-Petrareanu, G.; Visinescu, D.; Dinca, V.; Bonciu, A.; Mihailescu, C.N.; Purcarea, C.; Boukherroub, R.; Szunerits, S.; Vasilescu, A. Flow injection enzymatic biosensor for aldehydes based on a Meldola Blue-Ni complex electrochemical mediator. *Microchim. Acta* **2020**, *187*, 550. [CrossRef]

104. Vasilescu, A.; Titoiu, A.M.; Purcarea, C.; Necula-Petrareanu, G. Method for Determining the Fungicide Thiram Based on Enzymatic Inhibition and Electrochemical Sensor. Romanian OSIM Patent Application No. A/00587, 13 August 2018.

105. Nagy, I.; Schoofs, G.; Compernolle, F.; Proost, P.; Vanderleyden, J.; de Mot, R. Degradation of the thiocarbamate herbicide EPTC (S-ethyl dipropylcarbamothioate) and biosafening by *Rhodococcus* sp. strain NI86/21 involve an inducible cytochrome P-450 system and aldehyde dehydrogenase. *J. Bacteriol.* **1995**, *177*, 676–687. [CrossRef]

106. Bell, K.S.; Philp, J.C.; Aw, D.W.; Christofi, N. The genus *Rhodococcus*. *J. Appl. Microbiol.* **1998**, *85*, 195–210. [CrossRef] [PubMed]

107. De Carvalho, C.C.C.R.; da Fonseca, M.M.R. The remarkable *Rhodococcus* erythropolis. *Appl. Microbiol. Biotechnol.* **2005**, *67*, 715–726. [CrossRef] [PubMed]

108. Van der Geize, R.; Dijkhuizen, L. Harnessing the catabolic diversity of rhodococci for environmental and biotechnological applications. *Curr. Opin. Microbiol.* **2004**, *7*, 255–261. [CrossRef] [PubMed]

109. Singh, G.; Bhalla, A.; Kaur, P.; Capalash, N.; Sharma, P. Laccase from prokaryotes: A new source for an old enzyme. *Rev. Environ. Sci. Biotechnol.* **2011**, *10*, 309–326. [CrossRef]

110. Prakash, O.; Mahabare, K.; Yadav, K.K.; Sharma, R. Fungi from Extreme Environments: A Potential Source of Laccases Group of Extremozymes. In *Fungi in Extreme Environments: Ecological Role and Biotechnological Significance*; Tiquia-Arashiro, S.M., Grube, M., Eds.; Springer International Publishing: Cham, Switzerland; Amsterdam, The Netherlands, 2019; pp. 441–462.

111. Bains, J.; Capalash, N.; Sharma, P. Laccase from a non-melanogenic, alkalotolerant gamma-proteobacterium JB isolated from industrial wastewater drained soil. *Biotechnol. Lett.* **2003**, *25*, 1155–1159. [CrossRef]

112. Suzuki, T.; Endo, K.; Ito, M.; Tsujibo, H.; Miyamoto, K.; Inamori, Y. A thermostable laccase from *Streptomyces lavendulae* REN-7: Purification, characterization, nucleotide sequence, and expression. *Biosci. Biotechnol. Biochem.* **2003**, *67*, 2167–2175. [CrossRef]

113. Oliveira, T.M.B.F.; Ribeiro, F.W.P.; Sousa, C.P.; Salazar-Banda, G.R.; de Lima-Neto, P.; Correia, A.N.; Morais, S. Current overview and perspectives on carbon-based (bio)sensors for carbamate pesticides electroanalysis. *TrAC Trends Anal. Chem.* **2020**, *124*, 115779. [CrossRef]

114. Pichon, V.; Delaunay, N.; Combès, A. Sample Preparation Using Molecularly Imprinted Polymers. *Anal. Chem.* **2020**, *92*, 16–33. [CrossRef]

115. Amine, A.; Arduini, F.; Moscone, D.; Palleschi, G. Recent advances in biosensors based on enzyme inhibition. *Biosens. Bioelectron.* **2016**, *76*, 180–194. [CrossRef]

116. Del Valle, M. Bioelectronic Tongues Employing Electrochemical Biosensors. In *Trends in Bioelectroanalysis*; Matysik, F.-M., Ed.; Springer International Publishing: Berlin/Heidelberg, Germany, 2017; pp. 143–202.

117. Mishra, R.K.; Hubble, L.J.; Martín, A.; Kumar, R.; Barfidokht, A.; Kim, J.; Musameh, M.M.; Kyratzis, I.L.; Wang, J. Wearable Flexible and Stretchable Glove Biosensor for On-Site Detection of Organophosphorus Chemical Threats. *ACS Sens.* **2017**, *2*, 553–561. [CrossRef]

MDPI

St. Alban-Anlage 66

4052 Basel

Switzerland

Tel. +41 61 683 77 34

Fax +41 61 302 89 18

www.mdpi.com

Biosensors Editorial Office

E-mail: biosensors@mdpi.com

www.mdpi.com/journal/biosensors